Geophysical Monograph Series

Including

IUGG Volumes
Maurice Ewing Volumes
Mineral Physics Volumes

GEOPHYSICAL MONOGRAPH SERIES

Geophysical Monograph Volumes

1. **Antarctica in the International Geophysical Year** *A. P. Crary, L. M. Gould, E. O. Hulburt, Hugh Odishaw, and Waldo E. Smith (Eds.)*
2. **Geophysics and the IGY** *Hugh Odishaw and Stanley Ruttenberg (Eds.)*
3. **Atmospheric Chemistry of Chlorine and Sulfur Compounds** *James P. Lodge, Jr. (Ed.)*
4. **Contemporary Geodesy** *Charles A. Whitten and Kenneth H. Drummond (Eds.)*
5. **Physics of Precipitation** *Helmut Weickmann (Ed.)*
6. **The Crust of the Pacific Basin** *Gordon A. Macdonald and Hisashi Kuno (Eds.)*
7. **Antarctica Research: The Matthew Fontaine Maury Memorial Symposium** *H. Wexler, M. J. Rubin, and J. E. Caskey, Jr. (Eds.)*
8. **Terrestrial Heat Flow** *William H. K. Lee (Ed.)*
9. **Gravity Anomalies: Unsurveyed Areas** *Hyman Orlin (Ed.)*
10. **The Earth Beneath the Continents: A Volume of Geophysical Studies in Honor of Merle A. Tuve** *John S. Steinhart and T. Jefferson Smith (Eds.)*
11. **Isotope Techniques in the Hydrologic Cycle** *Glenn E. Stout (Ed.)*
12. **The Crust and Upper Mantle of the Pacific Area** *Leon Knopoff, Charles L. Drake, and Pembroke J. Hart (Eds.)*
13. **The Earth's Crust and Upper Mantle** *Pembroke J. Hart (Ed.)*
14. **The Structure and Physical Properties of the Earth's Crust** *John G. Heacock (Ed.)*
15. **The Use of Artificial Satellites for Geodesy** *Soren W. Henricksen, Armando Mancini, and Bernard H. Chovitz (Eds.)*
16. **Flow and Fracture of Rocks** *H. C. Heard, I. Y. Borg, N. L. Carter, and C. B. Raleigh (Eds.)*
17. **Man-Made Lakes: Their Problems and Environmental Effects** *William C. Ackermann, Gilbert F. White, and E. B. Worthington (Eds.)*
18. **The Upper Atmosphere in Motion: A Selection of Papers With Annotation** *C. O. Hines and Colleagues*
19. **The Geophysics of the Pacific Ocean Basin and Its Margin: A Volume in Honor of George P. Woollard** *George H. Sutton, Murli H. Manghnani, and Ralph Moberly (Eds.)*
20. **The Earth's Crust: Its Nature and Physical Properties** *John G. Heacock (Ed.)*
21. **Quantitative Modeling of Magnetospheric Processes** *W. P. Olson (Ed.)*
22. **Derivation, Meaning, and Use of Geomagnetic Indices** *P. N. Mayaud*
23. **The Tectonic and Geologic Evolution of Southeast Asian Seas and Islands** *Dennis E. Hayes (Ed.)*
24. **Mechanical Behavior of Crustal Rocks: The Handin Volume** *N. L. Carter, M. Friedman, J. M. Logan, and D. W. Stearns (Eds.)*
25. **Physics of Auroral Arc Formation** *S.-I. Akasofu and J. R. Kan (Eds.)*
26. **Heterogeneous Atmospheric Chemistry** *David R. Schryer (Ed.)*
27. **The Tectonic and Geologic Evolution of Southeast Asian Seas and Islands: Part 2** *Dennis E. Hayes (Ed.)*
28. **Magnetospheric Currents** *Thomas A. Potemra (Ed.)*
29. **Climate Processes and Climate Sensitivity (Maurice Ewing Volume 5)** *James E. Hansen and Taro Takahashi (Eds.)*
30. **Magnetic Reconnection in Space and Laboratory Plasmas** *Edward W. Hones, Jr. (Ed.)*
31. **Point Defects in Minerals (Mineral Physics Volume 1)** *Robert N. Schock (Ed.)*
32. **The Carbon Cycle and Atmospheric CO_2: Natural Variations Archean to Present** *E. T. Sundquist and W. S. Broecker (Eds.)*
33. **Greenland Ice Core: Geophysics, Geochemistry, and the Environment** *C. C. Langway, Jr., H. Oeschger, and W. Dansgaard (Eds.)*
34. **Collisionless Shocks in the Heliosphere: A Tutorial Review** *Robert G. Stone and Bruce T. Tsurutani (Eds.)*
35. **Collisionless Shocks in the Heliosphere: Reviews of Current Research** *Bruce T. Tsurutani and Robert G. Stone (Eds.)*
36. **Mineral and Rock Deformation: Laboratory Studies—The Paterson Volume** *B. E. Hobbs and H. C. Heard (Eds.)*
37. **Earthquake Source Mechanics (Maurice Ewing Volume 6)** *Shamita Das, John Boatwright, and Christopher H. Scholz (Eds.)*
38. **Ion Acceleration in the Magnetosphere and Ionosphere** *Tom Chang (Ed.)*
39. **High Pressure Research in Mineral Physics (Mineral Physics Volume 2)** *Murli H. Manghnani and Yasuhiko Syono (Eds.)*
40. **Gondwana Six: Structure, Tectonics, and Geophysics** *Gary D. McKenzie (Ed.)*

41 **Gondwana Six: Stratigraphy, Sedimentology, and Paleontoloty** *Garry D. McKenzie (Ed.)*

42 **Flow and Transport Through Unsaturated Fractured Rock** *Daniel D. Evans and Thomas J. Nicholson (Eds.)*

43 **Seamounts, Islands, and Atolls** *Barbara H. Keating, Patricia Fryer, Rodey Batiza, and George W. Boehlert (Eds.)*

44 **Modeling Magnetospheric Plasma** *T. E. Moore, J. H. Waite, Jr. (Eds.)*

45 **Perovskite: A Structure of Great Interest to Geophysics and Materials Science** *Alexandra Navrotsky and Donald J. Weidner (Eds.)*

46 **Structure and Dynamics of Earth's Deep Interior (IUGG Volume 1)** *D. E. Smylie and Raymond Hide (Eds.)*

47 **Hydrogeological Regimes and Their Subsurface Thermal Effects (IUGG Volume 2)** *Alan E. Beck, Grant Garvin and Lajos Stegena (Eds.)*

48 **Origin and Evolution of Sedimentary Basins and Their Energy and Mineral Resources (IUGG Volume 3)** *Raymond A. Price (Ed.)*

IUGG Volumes

1 **Structure and Dynamics of Earth's Deep Interior** *D. E. Smylie and Raymond Hide (Eds.)*

2 **Hydrogeological Regimes and Their Subsurface Thermal Effects** *Alan E. Beck, Grant Garvin and Lajos Stegena (Eds.)*

3 **Origin and Evolution of Sedimentary Basins and Their Energy and Mineral Resources** *Raymond A. Price (Ed.)*

Maurice Ewing Volumes

1 **Island Arcs, Deep Sea Trenches, and Back-Arc Basins** *Manik Talwani and Walter C. Pitman III (Eds.)*

2 **Deep Drilling Results in the Atlantic Ocean: Ocean Crust** *Manik Talwani, Christopher G. Harrison, and Dennis E. Hayes (Eds.)*

3 **Deep Drilling Results in the Atlantic Ocean: Continental Margins and Paleoenvironment** *Manik Talwani, William Hay, and William B. F. Ryan (Eds.)*

4 **Earthquake Prediction—An International Review** *David W. Simpson and Paul G. Richards (Eds.)*

5 **Climate Processes and Climate Sensitivity** *James E. Hansen and Taro Takahashi (Eds.)*

6 **Earthquake Source Mechanics** *Shamita Das, John Boatwright, and Christopher H. Scholz (Eds.)*

Mineral Physics Volumes

1 **Point Defects in Minerals** *Robert N. Schock (Ed.)*

2 **High Pressure Research in Mineral Physics** *Murli H. Manghnani and Yasuhiko Syono (Eds.)*

Slow Deformation and Transmission of Stress in the Earth

Steven C. Cohen
Petr Vaníček

Editors

American Geophysical Union
International Union of Geodesy and Geophysics

Geophysical Monograph/IUGG Series

Library of Congress Cataloging-in-Publication Data
Slow deformation and transmission of stress in the earth / edited by
Steven C. Cohen, Petr Vaníček.
p. cm. — (Geophysical) monograph; 49/IUGG series; 4)
  Includes bibliographies.
  ISBN 0-87590-453-X
  1. Rock deformation. 2. Plate tectonics.   I. Cohen, S.C.
(Steven C.) II. Vaníček, Petr. 1935–   . III. Series.
QE604.S57   1989
551.8—dc19   89-6456

Copyright 1989 by the American Geophysical Union, 2000 Florida Avenue, NW, Washington, DC 20009

Figures, tables, and short excerpts may be reprinted in scientific books and journals if the source is properly cited.

   Authorization to photocopy items for internal or personal use, or the internal or personal use of specific clients, is granted by the American Geophysical Union for libraries and other users registered with the Copyright Clearance Center (CCC) Transactional Reporting Service, provided that the base fee of $1.00 per copy, plus $0.10 per page is paid directly to CCC, 21 Congress Street, Salem, MA 01970. 0065-8448/89/$01. + .10.
   This consent does not extend to other kinds of copying, such as copying for creating new collective works or for resale. The reproduction of multiple copies and the use of full articles or the use of extracts, including figures and tables, for commercial purposes requires permission from AGU.

Printed in the United States of America.

CONTENTS

Preface
Steven C. Cohen and Petr Vaníček   ix

Introduction
Steven C. Cohen and Petr Vaníček   xi

1. **Post-Glacial Rebound Analysis for a Rotating Earth**
   Dazhong Han and John Wahr   1
2. **On the Figures of the Earth**
   R. Tonn, J. Zschau   7
3. **Contemporary Vertical Crustal Motion in the Pacific Northwest**
   Sandford R. Holdahl, Francois Faucher, and Herb Dragert   17
4. **Kinematics and Mechanics of Tectonic Block Rotations**
   Amos Nur, Hagai Ron, and Oona Scotti   31
5. **Plate Motions, Earth's Geoid Anomalies, and Mantle Convection**
   Fu Rong-shan   47
6. **Role of Episodic Creep in Global Mantle Deformation**
   G. Ranalli and H. H. Schloessin   55
7. **Layered Block Model in Problems of Slow Deformations of the Lithosphere and of Earthquake Engineering**
   A. D. Gvishiani, V. A. Gurvich, and A. G. Tumarkin   65
8. **Geodetic Measurement of Deformation East of the San Andreas Fault in Central California**
   Jeanne Sauber, Michael Lisowski and Sean C. Solomon   71
9. **Earthquakes' Impact on Changes in Height**
   I. Joó   87
10. **Vertical Movement of Indo-Gangetic Plains**
    C. S. Joshi, A. N. Singh, Manohar Lal   97
11. **Strain Analysis of Tectonic Movements From Geodetic Data Across Krol and Nahan Thrusts**
    C. S. Joshi, A. N. Singh and Atam Prakash   107
12. **Viscoelastic Deformations and Temporal Variations in the Geopotential**
    Roberto Sabadini, David A. Yuen and Paolo Gasperini   115
13. **Migration of Vertical Deformations and Coupling of Island Arc Plate and Subducting Plate**
    Satoshi Miura, Hiroshi Ishii and Akio Takagi   125

PREFACE

The papers presented in this monograph are based on presentations made at the symposium on "Slow Deformation and Transmission of Stress in the Earth" at the XIXth General Assembly Meeting of the International Union of Geodesy and Geophysics. The symposium was held on the campus of the University of British Columbia in Vancouver, Canada on August 15th and 17th 1987.

The editors would like to thank the authors, reviewers and AGU staff who contributed to the publication of this monograph.

Steven C. Cohen and Petr Vaníček

INTRODUCTION

A symposium entitled "Slow Deformation and Transmission of Stress in the Earth" was convened at the XIXth General Assembly Meeting of the International Union of Geodesy and Geophysics. This monograph is based on presentations made at that symposium which was held on the campus of the University of British Columbia in Vancouver, Canada on August 15th and 17th, 1987. The objective of the symposium was to engage geophysicists and geodesists in a discussion of the mechanisms, models, and measurements of slow deformations and stress transmission in the Earth's crust and mantle. These deformations are characterized by their quasi-static nature in which the effects of acceleration are negligible compared to those due to gravitational, rheological, thermal, chemical, and phase-change stresses. Phenomena such as tectonic plate motions, postglacial rebound, mantle convection, strain accumulation, aseismic strain release, and polar motion-induced deformations are included in this description. For many phenomena the effects of anelasticity and temporal nonlinearity are significant. This view of slow deformations is largely geophysical. From a geodetic point of view, however, the maintenance of accurate coordinates of points on the Earth's surface requires that the geometry of ongoing deformation be taken into account [International Association of Geodesy, 1987]. This necessitates an understanding of the geophysical models of the temporal deformations and their predictive powers. Thus the symposium presenters were challenged to address such key issues as: the current state of understanding of the phenomena of slow deformation and transmission of stress, contrasts between alternative models, applicability of geophysical models to prediction of positional changes, and assessment of model and parameter accuracies. These were formidable challenges and as such were pursued by only some of the authors.

The symposium consisted of approximately three dozen invited and contributed papers. About two-thirds of the papers were presented orally and the remainder as poster papers. The authors came from over a dozen countries including: Canada, China, Czechoslovakia, Egypt, France, Great Britain, Hungary, India, Italy, Japan, the Soviet Union, Sweden, and the United States. The first day of the symposium focused on geodetic measurement of crustal movements and models of the various aspects of the earthquake cycle. The second day covered a wider range of issues such as intracontinental deformation, thermoviscoelasticity, transient creep, Pleistocene deglaciation and magma solitons. Subsequent to the symposium, the presenters were invited to contribute papers based on their presentations for inclusion in this volume. About half of the authors chose to submit papers, others declined, usually on the grounds that their papers were being considered for publication elsewhere. The submitted papers were reviewed for technical content following the standard procedures of the American Geophysical Union and returned to the authors for revisions when required. A few of the papers were rejected or withdrawn during the review process. The remainder form this volume.

In this introduction we limit our overview to those papers which are presented in this volume. We begin with a discussion of the whole earth and perturbations to its gravity field due to surface loads.

Sabadini, et al. discuss the long wavelength gravity signatures due to present day and Pleistocene forcings. It has been recognized for some time that the viscoelastic response of the Earth to surface loading and unloading produces time-dependent signatures in the gravity field [e.g., Heiskanen and Vening Meinesz, 1958]. Recently, the analysis of the orbit of the Laser Geodynamics Satellite (LAGEOS) has resulted in determination of the rate of change of geopotential coefficients of the Earth's gravity field such as J_2 or C_{20}. These changes have been attributed to the viscoelastic rebound of the Earth to post-Pleistocene deglaciation. Changes in J_2 have been used by a number of researchers to constrain estimates of mantle viscosity [e.g., Peltier, 1985]. In the present paper the authors use a linear, transient viscoelastic rheological model of the Earth to show that the long wavelength components of the gravity field are sensitive to current glacial melting and changes in the Antarctic ice sheet mass. Their model results suggest that transient gravitational responses depend strongly on the ratio between short and long term viscosities.

Continuing on the theme of postglacial rebound, Han and Wahr discuss the effects of variations in the Earth's rotation on the centrifugal force and hence of the deformation of the Earth in response to this effect. They find that the effect of rotation on centrifugal force partially cancels that of postglacial rebound shifts in the Earth's inertia tensor. The analysis is then used to estimate the effects of Earth rotation on the determinations of lower mantle viscosity and lithospheric thickness from postglacial rebound models, free air gravity data, and relative sea level data. They find that the biggest effects occur in modeling the relative sea level near the edge of the region originally loaded by the sheet.

The rheological properties of the Earth determine the time scale of perturbations in response to applied stresses. While most convection models assume a steady-state creep law for mantle rocks [e.g., Richter, 1973], the possibility that time and spatially-dependent creep can influence postglacial rebound has been raised by several authors. Ranalli and Schloessin have considered several mechanisms of creep in the lower mantle including transient creep—in which strain rate is time-dependent, changes in creep due to pressure and temperature changes in the local environment, dynamic recrystallization, and several others possibilities. They summarize the conditions under which the various time-dependent creep mechanisms are likely to be significant for deformation in the mantle. Although they find that some of the creep mechanisms that they have considered are unlikely to be important under conditions found in the mantle, they conclude that spatial and temporal variations in mantle creep can be significant.

Tonn and Zschau have revisited the problem of the Earth's fossil bulge, the physical significance of which was much discussed a few decades ago by Ledersteger [1967], Goldreich and Toomre [1969], and others. They studied the problem by attempting to trace the evolution in time of the flattenings of the geoid, the hydrostatic figure of the Earth and the Earth itself. The hydrostatic figure flattening depends on relaxation in bulk which is as yet unknown. It appears, however, that the unrelaxed bulk modulus yields a value of flattening which is in good agreement with other published values. It will be interesting to see their results for the Earth (topographic) flattening and the conclusions they will draw from these.

The idea that the tectonic plate motions are part of a convective system in the mantle is one of the basic geodynamic tenets [Turcotte, 1975]. However, the development of dynamic models has proven to be a formidable problem involving the nonlinear coupling of mechanical, gravitational, thermal, and chemical effects. Consequently, numerical modelers have used developed simplified models to address various aspects of the problem. In order to study the relationship between the plate motions, mantle flow, and geoid undulations, Fu employs the Boussinesq approximation to the Navier-Stokes equation in which density perturbations are considered in the conservation of momentum equations only to the extent to which they affect the gravitational body force term. The model is used to determine the correlation coefficient between the poloidal components of the velocity of plate motions and the surface velocity field of convection.

There is, of course, a wide range of geophysical models which can be discussed under the general topic of slow deformations. For example, Gvishiani, et al. discuss a layered block model which can be applied to such problems as the coupled interaction of rigid lithospheric blocks and a deformable fault zone. They show how the principles of convex analysis can be used to determine the equilibrium positions of an interacting system of rigid and deformable blocks.

Nur, et al. focus on a more specific problem by reviewing the theory of microblock rotations produced when sets of parallel or subparallel, closely-spaced faults are subjected to shearing. Initially, the rotation of the blocks is accompanied by motion on the original faults. However, after a certain degree of rotation has occurred, further motion is more likely to result in the formation of new faults rather than slip on the original ones alone. Thus very larger rotations are accompanied by multiple fault sets such as those observed in the western United States and elsewhere.

The analysis of geodetic data takes many forms. In the simplest form, horizontal or vertical survey data yield relative site locations. The variation in these relative positions with time can often be correlated with tectonic phenomena such as the occurrence of an earthquake, the accumulation of strain, mountain building, etc. On a somewhat more detailed level, numerical models of both surface and subsurface deformations can be developed. Generally, these models use knowledge of local geological features to derive a structural representation of the region under study. Then the surface deformations are constrained by the observational data. Following this approach, Ishii et al. compare observations of vertical motion in northeastern Japan taken over a several decade time span with the results of finite element analyses. They conclude that the Pacific Ocean side of the island is subsiding and the Japanese Sea side is rising.

Relevelings and sea level rise as seen by tide gauges are the standard geometrical data used for studying crustal uplift and/or subsidence. By means of such geometrical data, Holdahl et al. found the uplift velocity for the coastal strip of the state of Washington and southwestern British Columbia, Canada range between -2mm and $+3$ mm per year. In addition, their analysis indicates a coseismic subsidence of 10 cm accompanying the 1946 Vancouver Island earthquake and a discernable increase in uplift rates in Vancouver Island over the past decade. The pattern of the uplift is consistent with hypothesized oblique subduction, normal subduction, and non-subductive convergence off the south, central, and northern parts of the island, respectively. The results seem to be congruent with the postulated subduction zone underlying the whole region.

Localized co-seismic vertical movement of several centimeters associated with the 1985 Lake Balaton, Hungary earthquake of magnitude 5.6 is reported by Joó. His study is based on first order levelings bracketing the seismic event; the movement is clearly detectable against the overall background subsidence of 0.2 mm per year pervading the area of interest. Particular attention is paid to the determination of the statistical significance level of the detected movement and its horizontal gradient.

Selected first order repeated levelings were also used by Joshi, et al. to compute the linear vertical crustal motion pattern in the Indo-Gangetic plains. From the analysis they deduced that several earthquakes, which have occurred during the past five decades in the area, had coseismic uplifts ranging from a few centimeters to over 20 cm. The overall patterns show uniform tilts apparently associated with tectonic processes. These tilts seem to suggest that the downwarping of the northern edge of the Indian plate is continuing.

In a separate study, Joshi, et al. analyzed a four times observed, small, 150 m by 150 m, triangulation network covering a tectonically interesting patch divided by two closely spaced parallel trust faults in the Kalawar area of Himachal Pradesh state, India. They have estimated rigid block translations and homogeneous strain within the blocks. Sizeable changes in the dilatation of the blocks—the only reliably determinable parameter—occurred between 1973 and 1986. The estimated direction of maximum shear is compatible with the geometry of the faults.

Sauber, et al. use triangulation and trilateration data to estimate the shear strain rates in the Diablo Range of Central California and to estimate slip rates along the Calaveras and Paicines faults. Angle changes not involving stations located close to the Paicines fault indicate shortening at a rate 5.7 ± 2.7 mm/yr over a 30 km wide zone. The orientation of this shortening is consistent with the orientation of the major fold structures in the region. Further south and east, near Coalinga, the estimated rate of right-lateral slip on the Calaveras-Paicines fault system is 10 mm/yr.

The papers presented in this volume, the other papers presented at the IUGG symposium, and the general scientific literature indicate that geodetic data are providing many important constraints on geophysical models. Time-dependent position data are particularly important for studying the earthquake cycle, and both position and gravitational data have contributed to studies of postglacial rebound. Insight into mantle convection and tectonic processes has been advanced using various types of geodetic observations. Less progress has been made in applying geophysical models to the systematic prediction of geodetic position changes or correction to geodetic observations. In the future, geophysical models will evolve based on a mixture of geodetic, geophysical (seismic, magnetic, heat flow) and geological data. Model developments and data anlaysis will be greatly aided by increased computational power. With increasing frequency, the models required to explain both the geodetic and nongeodetic measurements, at their observational accuracies, will have to consider the coupling of mechanical and thermal processes and take into account the structural and geometric complexity of individual tectonic environments.

References

Goldreich, P., and A. Toomre, "Some Remarks on Polar Wandering," *J. Geophys. Res.*, 74, 2555–2567. 1969.

Heiskanen, W.A. and F.A. Vening Meinesz, "The Earth and its Gravity Field," Mc-Graw-Hill, 1958.

International Association of Geodesy, Special Study Group 4.96, Four-Dimensional Geodetic Positioning" P. Vaníček, editor, *Manuscripta Geodaetica*, 12(3), 147–222, 1987.

Ledersteger, K., "The Equilibrium Figure of the Earth and the Normal Spheriod," in *Analysed Proceeding of the International Symposium on the Figure of the Earth and Refraction*, K. Ledersteger, editor, pp. 20–22, Austrian Geodetic Commission, Vienna, Austria, March, 1967.

Peltier, W.R., "The LAGEOS Constraint on Deep Mantle Viscosity: Results From a New Normal Mode Method for the Inversion of Viscoelastic Relaxation Spectra," *J. Geophys. Res.* 90, 9411–9421, 1985.

Richter, F.M., "Dynamical Models for Sea Floor Spreading," *Rev. Geophys. Space Physics*, 11, 223–287, 1973.

Turcotte, D.L., "The Driving Mechanism of Plate Tectonics," *Rev. Geophys. Space Physics*, 13, 589–590, 1975.

Steven C. Cohen
Geodynamics Branch
Laboratory for Terrestrial Physics
Goddard Space Flight Center
Greenbelt, MD 20771

Petr Vaníček
Department of Surveying Engineering
University of New Brunswick
Fredericton, New Brunswick
Canada

POST-GLACIAL REBOUND ANALYSIS FOR A ROTATING EARTH

Dazhong Han and John Wahr

Cooperative Institute for Research in Environmental Sciences
University of Colorado, Boulder, CO 80309, U.S.A.

Abstract. The effects of the earth's rotation on post-glacial rebound are discussed. Those effects are best summarized as the result of a two part process. First, post-glacial rebound induces changes in the earth's inertia tensor which lead to changes in the earth's rotation vector. These effects on the rotation vector have been considered in previous studies. Second, the variability in rotation perturbs the centrifugal force, and causes additional deformation of the earth which affects such things as relative sea level and free air gravity. These additional deformation effects have not been considered before. We find that these effects are potentially important when modelling relative sea level at places close to the edge of the original ice sheet, but are small for the free air gravity anomaly at the center of rebound.

Introduction

Recent post-glacial rebound studies (see, for example, Wu and Peltier, 1982, 1983, 1984; Yuen et al, 1986) have used observations of relative sea level changes and of free air gravity anomalies to learn about mantle viscosity. The results, which assume a Maxwell rheology for the mantle, suggest that the average mantle viscosity is about 10^{21} Pa s, and that there is little significant increase in viscosity from upper to lower mantle. This conclusion appears to conflict with results from Richards and Hager (1984), who were able to model much of the earth's long wavelength geoid anomalies as caused by internal forcing of the earth by assuming a contrast of 2 orders of magnitude between the upper and lower mantle viscosities. This inconsistency between the post-glacial rebound results and the results of Richards and Hager (the latter further supported by laboratory evidence - see Sammis et al, 1977) led Sabadini et al (1985) and Peltier et al (1986) to suggest that both results could be accommodated by assuming a more general mantle rheology.

However, existing models of post-glacial deformation do not include the effects of the earth's rotation in a consistent manner. Post-glacial deformation perturbs the earth's inertia tensor, which leads to variations in the earth's rotation vector. Those variations are included in existing models. The variable rotation causes perturbations in the centrifugal force which further deform the earth and affect observations of free air gravity, relative sea level, etc. Those deformation effects have not been considered previously. This led us to wonder whether the inclusion of those effects could help to resolve the discrepancy between the post-glacial and model-geoid viscosity estimates, without requiring a change in the rheological model.

To address this problem, we modelled the deformation caused by the incremental centrifugal force, with results described below. We find that the rotational deformation is almost entirely described with a $Y_2^{\pm 1}$ spherical harmonic angular dependence, and that it tends to cancel the $Y_2^{\pm 1}$ component for a non-rotating earth in both relative sea level and free air gravity. The effect of the rotational deformation on relative sea level at places close to the edge of the original ice sheet is potentially important. However, the effect on the gravity anomaly at the center of rebound is small. We find that the inclusion of rotation does not appreciably affect the post-glacial viscosity constraints, but can be important when estimating lithospheric thicknesses from relative sea level observations.

A Review of the Theory of Glacial Isostasy

Two important types of data related to post-glacial rebound are particularly useful for constraining mantle viscosity: relative sea level histories and the present day negative gravity anomaly over the main center of rebound. These and other observable quantities are most conveniently modelled by convolving models of the ice and water load with viscoelastic Green's functions. For example, if $\xi(\theta,\lambda,t)$ represents the change in sea level height at co-latitude θ, eastward longitude λ, and time t, then

$$\xi(\theta,\lambda,t) = \left[\int L(\theta',\lambda',t')\, G_\xi(\theta',\lambda',t';\theta,\lambda,t)\, \sin\theta' d\theta' d\lambda' dt' + D\right]\frac{C(\theta,\lambda)}{g} \quad (1)$$

where $C(\theta,\lambda)$ is the ocean function, $L(\theta,\lambda,t)=L_I+L_o$ is the total applied load at the position (θ,λ) at time t (L_I represents the ice load and L_o is the ocean load resulting from the addition of melted ice to the oceans), and D is a constant required to conserve in the ocean.

The Green's function $G_\xi(\theta',\lambda',t';\theta,\lambda,t)$ represents the contribution to ξ at position (θ,λ) and time t from a delta func-

Copyright 1989 by
International Union of Geodesy and Geophysics
and American Geophysical Union.

tion load at position (θ',λ') and time t'. G_ξ is computed by expanding the delta function spatial dependence of the load as a sum of spherical harmonics, finding the response of the earth to each spherical harmonic, and then adding those responses together. For a spherical, non-rotating earth, a Y_n^m load induces, exactly, a Y_n^m response. For the real earth, non-spherical effects (including the effects of rotation) are small enough that the assumption of a Y_n^m response to a Y_n^m load is good to 1% or better at long wavelengths. So, the response to a Y_n^m load can be described with scalar Love numbers (actually, functions of time) representing the Y_n^m deformation at time t caused by a Y_n^m spherical harmonic load with a delta function time dependence at t=0. Let $k_n^m(t)$ and $h_n^m(t)$ be the Love numbers representing the gravitational potential and the vertical displacement of the surface, respectively. Then G_ξ can be expanded as

$$G_\xi(\theta',\lambda',t';\theta,\lambda,t) = \frac{4\pi a}{M_e} \sum_{n,m} \left[-h_n^m(t-t') \right. \quad (2)$$

$$\left. + k_n^m(t-t') \right] \frac{1}{2n+1} Y_n^m(\theta,\lambda) \overline{Y_n^m(\theta',\lambda')}$$

where M_e is the mass of the earth, the Y_n^m are complex spherical harmonics (normalized so that the integral of $|Y_n^m|^2$ over the unit sphere is unity), and the bar over Y_n^m represents complex conjugation. Green's functions for other observables, such as the free air gravity anomaly, are similar.

(For a spherically symmetric, non-rotating earth, h_n^m and k_n^m are independent of m, and G_ξ reduces to the more familiar result:

$$G_\xi(\theta',\lambda',t';\theta,\lambda,t) = \frac{a}{M_e} \sum_{n=0}^{\infty} (-h_n(t-t') + k_n(t-t')) P_n(\cos\alpha) \quad (3)$$

where the superscript m on the Love numbers has been dropped, and α is the angle between the surface points at (θ,λ) and (θ',λ'). We will find, though, that for a rotating earth the n=2 Love numbers do depend on m, and so (2) does not reduce to (3).)

The Love numbers can be written in the form (Peltier, 1974):

$$h_n^m = h^E + \sum_{j=1}^{M} r_j^h e^{-s_j t} \quad (4)$$

$$k_n^m = k^E + \sum_{j=1}^{M} r_j^k e^{-s_j t} \quad (5)$$

where h^E and k^E represent the instantaneous "elastic" part of the earth's response, and the summation terms in (4) and (5) describe the earth's viscous relaxation. The summations are over a discrete set of relaxation times ($\tau_j = 1/s_j$) where the s_j are eigenvalues of the Laplace transformed field equations for the surface loading problem.

The Effects of the Earth's Rotation

The Love numbers $h_2^{\pm 1}$ and $k_2^{\pm 1}$ are affected by the earth's rotation. This is best understood as follows. Suppose that

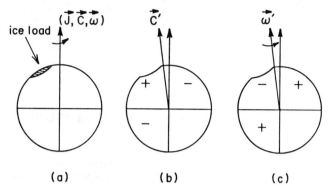

Figure 1. A pictorial representation of the rotation effect (see text).

prior to the removal of the ice load, the earth is in an equilibrium state as shown in Figure 1a, where the earth's figure axis $\vec{C}$, angular momentum axis $\vec{J}$, and rotation axis $\vec{\omega}$ are coincident. When the ice load melts, the removal of mass causes a shift of the figure axis from $\vec{C}$ to $\vec{C}'$ as shown in Figure 1b. Also shown in Figure 1b is the pattern of deformation caused by the removal of the load: '+' represents uplift, and '-' represents subsidence. To conserve angular momentum, the earth's rotation axis must shift in the same direction as the figure axis relative to the earth's surface (actually, the figure and rotation axes remain fixed in inertial space, it is the earth that tips). The new rotation axis is represented by $\vec{\omega}'$ in Figure 1c. The new position of the rotation axis causes a reconfiguration of the centrifugal force in the earth, and the resulting change in centrifugal potential can be represented as a $Y_2^{\pm 1}$ potential, as shown below. This, in turn, induces $Y_2^{\pm 1}$ terms in the deformation, and so modifies the n=2, m=± 1 Love numbers. The effect of these rotation terms is to reduce the Love numbers, as can be inferred by comparing the deformation pattern caused by the change in centrifugal force (Figure 1c) with the pattern shown in Figure 1b. In fact, we find that the rotation-induced deformation tends to cancel the non-rotational contributions to the n=2, m=± 1 Love numbers.

To find the effects of rotation, consider the loading problem in the Laplace transformed domain. Let the Laplace transformed angular velocity vector of the earth be

$$\omega = \Omega(m_1, m_2, 1+m_3) \quad (6)$$

were $\Omega = \Omega \hat{z}$ is the angular velocity vector before removal of the load, and the m_i describe the effects of the melting/rebound process. Let I_{ij} represent the change in the earth's inertia tensor due to the combined effects of the load, the load-induced deformation, and the deformation caused by the change in rotation. Define the complex parameters $m_\pm = m_1 \pm i m_2$ and $I_\pm = I_{13} \pm i I_{23}$. Using the equations for angular momentum conservation (Lambeck, 1980) and assuming periods much longer than the free precessional period of the earth (14 months), $m_\pm$ and m_3 can be related to $I_\pm$ and I_{33} as:

$$m_\pm = \frac{I_\pm}{C-A} \quad (7)$$

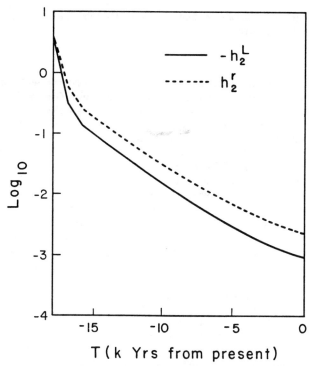

Figure 2. The logarithm, in the time domain, of the negative time domain Love number for a non-rotating earth, h_2^L (solid curve), and the contribution from rotation, h_2^r (dotted line). The earth model includes a 120 km thick elastic lithosphere. The two terms tend to cancel.

$$m_3 = -\frac{I_{33}}{C} \quad (8)$$

C and A in (7) and (8) are the unperturbed principal moments of inertia along the polar and equatorial axes, respectively. Since $(C-A)/C \approx 1/300$, and I_{33} is the same order as $I_\pm$, m_3 is roughly 300 times smaller than $m_\pm$. Thus, the deformation caused by the change in rotation rate (m_3) is much smaller than that caused by the change in the position of the axis ($m_\pm$), and will be ignored here.

Suppose a Laplace transformed, n=2, m=1 load is applied to the earth. Let the surface gravitational potential from this load be $\phi = \phi_2^1 Y_2^1$. The deformation induced by the load will lead to an additional surface gravitational potential $\phi^L = k_2^L \phi$, where k_2^L is the Laplace transformed n=2 load Love number for a spherical, non-rotating earth. MacCallagh's formula, which relates the inertia tensor to n=2 spherical harmonic components of the earth's gravitational potential (Lambeck, 1980, equation (2.4.5)), implies that

$$I_- = \sqrt{5/6\pi}\,\frac{a^3}{G}\,\phi_2^1(1+k_2^L) + I_-^r \quad (9)$$

where I_-^r represents the effects of the rotation-induced deformation. The load and load-induced deformation leads to a non-zero m_- through equation (7). (Note that since there is no load contribution to I_+, m_+ is zero.)

First, though, we relate I_-^r to m_-. The first order change in the centrifugal potential caused by m_- is

$$\psi = \sqrt{2\pi/15}\,\Omega^2 a^2 m_- Y_2^1 \quad (10)$$

ψ induces deformation in the earth which results in an additional gravitational potential given by $\psi_D = k_2^B \psi$, where k_2^B is the Laplace transformed n=2 body tide Love number for a spherical, non-rotating earth. ($\psi_D = k_2^B \psi$ is valid in the Laplace transformed, domain but not in the time domain. In the time domain, multiplication by k_2^B is replaced by convolution over time.) From MacCallagh's formula, we relate I_-^r to ψ_D and find (using (10))

$$I_-^r = \frac{a^5 \Omega^2}{3G}\,k_2^B m_- \quad (11)$$

Equations (11), (9), and (7) can then be combined to give a linear equation for m_- in terms of ϕ_2^1. This equation is easily solved, and the result for m_- is used in (10) to give the centrifugal potential in terms of the applied load, ϕ_2^1. We find:

$$\psi = \frac{1+k_2^L}{k_f - k_2^B}\,\phi_2^1 Y_2^1 \quad (12)$$

where k_f is the fluid Love number: $k_f = 3G(C-A)/a^5\Omega^2$.

Both ψ and ϕ cause deformation in the earth. For example, the total change in the gravitational potential due to

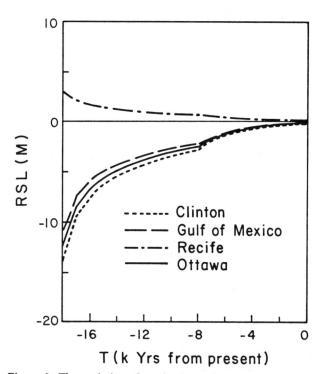

Figure 3. The variation of sea level at selected points due to the change in the earth's rotation (the contributions from k_2^r and h_2^r). Recife is in Brazil and Clinton is in Massachusetts. The lithospheric thickness is assumed to be 120 km. Note that the cumulative effect since the beginning of the last melting event can be larger than 10 meters.

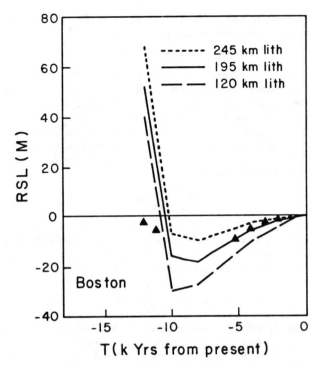

Figure 4. (Taken from Peltier, 1984) Comparison of the predicted and observed relative sea level history at Boston. The dashed, solid, and dotted curves are predictions for the models that have lithospheric thicknesses of 120, 195 and 245 km, respectively. The results are for a lower mantle viscosity of 10^{21} Pa s, and do not include the effects of rotation. The solid lozenges represent the data.

deformation inside the earth is k_2^B times ψ (ψ given by (12)), plus $k_2^L \phi_2^1 Y_2^1$. The result can be written as $k_2^1 \phi_2^1 Y_2^1$, where

$$k_2^1 = k_2^L + k_2^r = k_2^L + k_2^B \left(\frac{1+k_2^L}{k_f - k_2^B} \right) \qquad (13)$$

can be identified as the Laplace transformed, n=2, m=1 Love number for a rotating earth. Similarly, the Laplace transformed vertical displacement Love number is

$$h_2^1 = h_2^L + h_2^r = h_2^L + h_2^B \left(\frac{1+k_2^L}{k_f - k_2^B} \right) \qquad (14)$$

where h_2^L and h_2^B are the Laplace transformed, n=2 load and body tide displacement Love numbers for a spherical, non-rotating earth. k_2^r and h_2^r in (13) and (14) represent the effects of rotation on the n=2, m=1 Love numbers. The n=2, m=−1 Love numbers are similarly perturbed.

The results (13) and (14) describe the Love numbers in the Laplace transformed domain. When these Love numbers are transformed to the time domain, they can be written in the form described by equations (4) and (5). The time domain Love numbers are then used in (2) to find G_ξ (and in

similar equations to find the other Green's functions) for a rotating earth.

Results

Previous studies have not included the rotational terms, k_2^r and h_2^r, when constructing G_ξ and the other Green's functions. The objective of our work is to assess the effects of those terms. We use an earth model with an elastic lithosphere of variable thickness, a uniform viscosity of 10^{21} Pa s throughout the rest of the mantle, and elastic parameters from model 1066B of Gilbert and Dziewonski (1975). We find k_2^r and h_2^r in the time domain using the method described by Peltier and Wu (1982). Briefly: (1) the Laplace transformed equations of motion are used to find the eigenvalues (s_j) for k_2^r and h_2^r (k_2^r, h_2^r, k_2^L, and h_2^L all have the same eigenvalues); (2) k_2^r and h_2^r are then expanded in the time domain as sums of $e^{-s_j t}$ as shown in (4) and (5), and the r_j^k and r_j^h are determined by least squares fitting in the Laplace transformed domain.

The effects of rotation on relative sea level and on free air gravity are computed for a disc ice load of radius $\alpha = 15°$ and a sawtooth wave history, by using k_2^r and h_2^r in (2) (and in the similar equation for the gravity Green's function) and convolving with the load using (1). The melted ice is assumed to be uniformly re-distributed over the ocean. We find that the effects of the earth's rotation tend to cancel the n=2, m=1 results for the non-rotating earth. (see, for example, Figure 2). It is not hard to show, in fact, that the rotational and non-rotational contributions cancel exactly for a homogeneous,

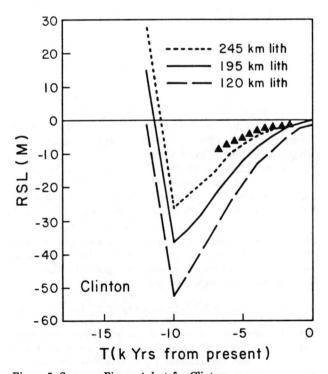

Figure 5. Same as Figure 4, but for Clinton.

incompressible earth. The rotation correction for the gravity anomaly at the center of the disc is small, however: only 1% of the total gravity anomaly there. This is because the Y_2^1 component for a non-rotating earth has only about a 1% effect on that anomaly to begin with. On the other hand, for relative sea level the rotation effect at some sites can be significant. The rotation correction for sea level is negative in the northern hemisphere and positive in the southern hemisphere (consistent with the spatial dependence of Y_2^1), and can be larger than 10 meters at times long after the melting. Figure 3 shows the rotation effect on sea level at selected locations. The largest corrections occur at mid-latitudes, including at points near the edge of the original Laurentide ice sheet.

As pointed out by Peltier (1984) and Yuen et al. (1986), the non-monotonic time histories of sea level at the edge of the ice sheet (for example at Boston and Clinton, Massachusetts) provide important constraints on the rheological model, particularly on the thickness of the elastic lithosphere. For example, Peltier (1984) found that he needed a surprisingly thick lithosphere to fit the sea level data at Boston and Clinton (thicknesses of 195 km and 245 km, respectively). His results are shown in Figures 4 and 5, where the model results are for a non-rotating earth with a lower mantle viscosity of 10^{21} Pa s. The inclusion of the rotation effect tends to require even a thicker lithosphere in these regions (see Figures 6 and 7).

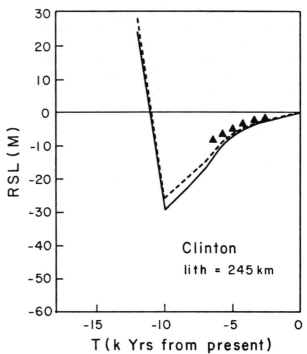

Figure 7. Same as Figure 6, but for Clinton and with a lithospheric thickness of 245 km.

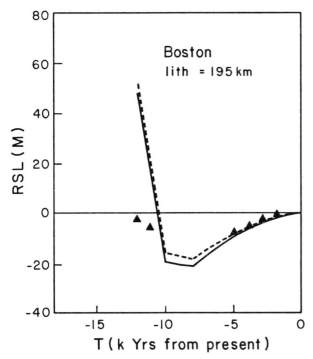

Figure 6. The predicted sea level history at Boston for a 195 km thick lithosphere, with (solid line) and without (dashed line) the rotational contributions. The solid lozenges represent the data. The addition of the rotational contributions degrades the fit to the data, and requires an even thicker lithosphere.

Conclusions

1) The inclusion of the earth's rotation in post-glacial rebound models leads to the near cancellation of the $Y_2^{\pm 1}$ components in both the free air gravity anomaly and relative sea level histories. The correction for free air gravity at the center of rebound is small. The maximum correction for relative sea level occurs at mid-latitudes and can be larger than 10 meters. However, both of these corrections are too small to have an appreciable effect on the solution for lower mantle viscosity.

2) The inclusion of rotation in the relative sea level calculations leads to estimates of lithospheric thickness which are even greater than the result obtained for a non-rotating earth.

<u>Acknowledgments.</u> We are grateful to David Yuen and Ben Chao for helpful discussion and comments. This work was supported in part by NASA grant NAG5-485.

References

Gilbert, F. & Dziewonski, A. M., An application of normal mode theory to the retrieval of structural parameters and source mechanisms from seismic spectra, *Phil. Trans. R. Soc. Lond. A, 278*, 187-269, 1985.

Lambeck, K., The Earth's Variable Rotation: Geophysical Causes and Consequences, *Cambridge University Press*, 449 pp, 1980.

Peltier, W. R., The impulse response of a Maxwell Earth, *Review of Geophysics and Space Physics, 12*, 649-669, 1974.

Peltier, W. R., The thickness of the continental lithosphere, *J. Geophys. Res., 89*, 11303-11316, 1984.

Peltier, W. R., R. A. Drummond, and A. M. Tushingham, Post-glacial rebound and transient lower mantle rheology, *Geophys. J. R. astr. Soc. 87*, 79-116, 1986.

Richards, M. A., and B. H. Hager, Geoid anomalies in a dynamic Earth, *J. Geophys. Res., 89*, 5987-6002, 1984.

Sabadini, R., and W. R. Peltier, Pleistocene de-glaciation and the Earth's rotation: implication for mantle viscosity, *Geophys. J. R. astr. Soc. 66*, 552-578, 1984.

Sabadini, R., D. A. Yuen, and P. Casprini, The effect of transient rheology on the interpretation of lower mantle viscosity, *Geophys. Res. Lett., 12*, 361-364, 1985a.

Sammis, C. G., J. C. Smith, G. Schubert, and D. A. Yuen, Viscosity-depth profile of polymorphic phase transitions, *J. Geophys. Res., 82*, 3747-3761, 1977.

Wu, P., and W. R. Peltier, Viscous gravitational relation, *Geophys. J. R. astr. Soc. 70*, 435-485, 1982.

Wu, P., and W. R. Peltier, Glacial isostatic adjustment and free air gravity anomaly as a constraint on deep mantle viscosity, *Geophys. J. R. astr. Soc., 74*, 377-449, 1983.

Wu, P., and W. R. Peltier, Pleistocene deglaciation and Earth's rotation: a new analysis, *Geophys. J. R. astr. Soc. 76*, 753-791, 1984.

Yuen, D. A., R. Sabadini, P. Gasperini, and E. Boschi, On transient rheology and glacial isostasy, *J. Geophys. Res., 91*, 11,420-11,438, 1986.

ON THE FIGURES OF THE EARTH

R. Tonn, J. Zschau

Institute of Geophysics, University of Kiel
D-2300 Kiel, West Germany

Abstract. The viscoelastic response of the Earth due to centrifugal force and the question as to whether or not an equilibrium figure exists have been investigated. In particular, we have studied the possible existence of a fossil equatorial bulge and the reason for the difference between the measured Geoid and the commonly adopted hydrostatic figure. By utilizing theoretical Love numbers a theory for the computation of the Earth's figure has been outlined. The investigation of a fluid Earth renders the computation of the hydrostatic figure (Hydroid) possible. We have shown that this figure cannot be determined unequivocally, because it is dependent on the unknown degree of relaxation in bulk. Consideration of the special case of a fluid Earth with unrelaxed bulk modulus yields a hydrostatic figure, which is in excellent agreement with published results from other authors. Using the same methods, but by using a Maxwell rheology for shear, and again assuming no relaxation of the bulk modulus, we have computed the spheroidal equipotential surface (Geoid) and the corresponding topographic surface (Topoid) as a function of time. The flattenings of the computed Geoid and Hydroid are equal, which implies that a fossil bulge cannot exist. However, the computed Geoid's flattening proved to be different from the measured Geoid's flattening. This suggests that either internal dynamic forces or relaxation in bulk are important. In the latter case, the Earth could well be in hydrostatic equilibrium. The difference between the measured Geoid and the theoretical Geoid cannot be taken as an indication for the Earth's deviation from hydrostatic equilibrium. However, no further information about relaxation in bulk is available, thus leaving the consideration of relaxation in bulk a goal for future investigations.

Finally, a new unequivocal criterion, which can be used to examine whether or not the Earth is in equilibrium, has been introduced. This requires that the shapes of the Topoid and Geoid be identical in the case of hydrostatic equilibrium. This comparison has not been able to be carried out so far due to a lack of data for the Topoid.

Introduction

The computation of the Earth's hydrostatic figure (Hydroid) is an old problem. Since 1743 and 1825, when Clairaut and Laplace published the results of their pioneering work, many scientists have been engaged in the computation of the hydrostatic figure and in its comparison with the measured figure of the Earth.

The literature concerned with the calculation of the hydrostatic figure falls into two categories: According to the internal theory, the hydrostatic figure is determined from a given density model and some gross parameters, such as the gravity constant and the Earth's angular velocity (see Zharkov and Trubitsyn, 1978; Lanzano, 1982; Denis, 1986, among others). The hydrostatic shape can, according to the external theory, also be determined by incorporating some observed parameters of the real Earth into the computation. Some of these are the dynamic form factor J_2 (Jeffreys, 1963), the dynamic ellipticity H (e.g., Khan, 1969), the ratio J_2/H (e.g., Nakiboglu, 1979), the trace of the inertia tensor TR(I) (e.g., Nakiboglu, 1982), or the principal moments of inertia (e.g., Khan 1969). In the external theory these quantities are assumed to be the same for the Geoid and the hydrostatic figure.

In principal, both methods are based on the general equation of equilibrium for a rotating fluid (see Budo, 1978, for example)

$$\rho \nabla U = \nabla P \qquad (1)$$

(∇: Nabla operator, P:pressure, ρ:density, U:potential).
Applying the curl operator (see, e.g., Zharkov and Trubitsyn, 1978) yields:

$$\nabla \rho \times \nabla U = 0 \qquad (2)$$

TABLE 1.

Hydrostatic flattening obtained by external theory

Reference		invariant parameter	inverse flattening
Jeffreys	(1963)	J_2	299.67 ± 0.05
Khan	(1969)	J_2	298.29 ± 0.05
Khan	(1969)	H	297.29 ± 0.05
Khan	(1969)	C	299.75 ± 0.05
Nakiboglu	(1979)	J_2/H	299.829
Denis	(1981)	J_2	298.88
Nakiboglu	(1982)	Trace I	299.638
Denis	(1983)	J_2	299.73

Hydrostatic flattening obtained by internal theory

Reference		earthmodel	inverse flattening
James & Kopal	(1963)	Bullen A	296.8
Nakiboglu	(1982)	PEM-A	299.627
Lanzano	(1982)	Bullen A	297.2
Lanzano	(1982)	QM 2	299.8
Lanzano	(1982)	1066 A	299.6
Lanzano	(1982)	HB 1	299.6
Denis	(1983)	PREM	299.68
Denis	(1986)	PEM-C	299.72
Denis	(1986)	PEM-O	299.7
Denis	(1986)	PEM-A	299.71

Hydrostatic flattening obtained by theoretical Love numbers

Reference		earthmodel	inverse flattening
Tonn	(this paper)	PREM	299.71

i.e., surfaces of equal density have to be surfaces of equal potential. This is the generally accepted condition for hydrostatic equilibrium. It describes the Earth in a fluid state without internal shear stresses, that is, with shear moduli completely relaxed to zero.

Table 1 gives an overview of corresponding results in the literature. A significant difference between the flattening of the computed Hydroids and the measured Geoid, as described by the Geodetic Reference System (GRS80) 1:298.257 (Moritz, 1980), is obvious. This difference is commonly interpreted as a deviation of the Earth from hydrostatic equilibrium. It is attributed either to a delay of the Earth's adjustment to slowly varying centrifugal forces, which results in the existence of a fossil equatorial bulge, to dynamic forces within the Earth's interior other than the centrifugal ones, or to both of these.

In this paper the above commonly adopted conclusions will be examined in detail. In particular, three questions are addressed.
1) Is the fossil equatorial bulge existent and can it at least partly be responsible for the observed difference between the Geoid and the Hydroid?
2. Is it justified to interpret the observed difference between the measured Geoid and the commonly adopted Hydroid as the Earth's deviation from hydrostatic equilibrium, and can the same observation be taken as an indication for the existence of internal dynamic forces, provided a fossil bulge has no significance?
3. Is there any unequivocal criterion that allows to decide whether the Earth is in hydrostatic equilibirum or not?

In order to answer these questions, we have split our study into three parts. In the first part we provide a new technique for the computation of the Earth's spheroidal figures, the hydrostatic one (Hydroid), the topographic one (Topoid), and the equipotential surface (Geoid). They are determined from computing the Earth's yielding to the slowly varying centrifugal force of its own rotation, involving the calculation of time dependent Love numbers for a viscoelastic Earth. As these figures should be different from each other, if the Earth were able to support a fossil bulge, their comparison will allow to give answer to the first question.

In the second part we will point out an important ambiguity in the determination of the Earth's hydrostatic figure, which causes a basic difficulty for deciding whether the Earth is in hydrostatic equilibrium or not. In fact, it will be shown that the common procedure of deciding this question from the observed difference between the Geoid and the adopted hydrostatic figure is not justified.

In the third part, we, therefore, suggest a new approach. It is based on comparing the shape of the Geoid with that of the Topoid, and it yields an unequivocal criterion which is independent of any theoretical computation.

Theoretical Estimation of the Geoid, Topoid, and Hydroid

Love Numbers

The formulas for the computation of the Earth's elastic deformation caused by periodic forces are well-known. The deformation can be expressed in terms of the constant parameters h and k, which describe the radial displacements of the solid Earth and its equipotential surface, respectively. These parameters were introduced by A.E.H. Love in 1909 and are called Love numbers. By utilizing the general linearized equation of motion for a self-gravitating elastic body (Phinney and Burridge, 1973) which is

$$-\rho\omega^2 \vec{u} = \nabla \cdot \overleftrightarrow{\tau} + \rho g \vec{e}_r (\nabla \cdot \vec{u}) - \rho \nabla(\psi + g u_r)$$
$$\nabla^2 \psi = -4\pi G \nabla \cdot (\rho \vec{u}) \quad (3)$$
$$\overleftrightarrow{\tau} = \lambda \overleftrightarrow{E} \nabla \cdot \vec{u} + 2\mu \{\nabla \vec{u} + (\nabla \vec{u})^T\}$$

(ω: angular velocity, u: displacement vector; u_r: radial component of the displacement vector; g: gravity, λ, μ: Lame's constants, E: unit matrix; T: transposed) and applying the correspondence principle, which allows the introduction of viscoelasticity, frequency dependent Love numbers can be computed (Zschau, 1979). In this study we have performed the calculations for a spherical and radially stratified Earth model. The model used here is the Preliminary Reference Earth Model (PREM) (Dziewonski and Anderson, 1981).

In the case of an elastic Earth the Love numbers are frequency dependent. In the case of a fluid Earth ($\mu=0$, μ: shear modulus) with static (d/dt=0) deformations, the equation of motion can be simplified to the equilibrium equation (1). For a viscoelastic Earth relaxation of the elastic moduli and their dispersion become important. This results in frequency and time dependence, respectively, of the Love numbers. For the Earth yielding to the slowly varying centrifugal forces either the fluid case or a steady state rheology such as the Maxwell Rheology has to be applied (the Maxwell Rheology is also applied to deformations with a much shorter time constant such as post glacial uplift (Peltier, 1974)). For our calculations we have used a Maxwell Rheology for shear, and we have assumed the bulk modulus not to relax. The fluid case was approximated by setting the shear modulus to zero. Several shear viscosity models were investigated. Mantle viscosities between 10^{22} and 10^{24} P were taken from Peltier (1974), Hager and O'Connell (1979), and Cathles (1975). Crustal viscosities were taken from Vetter (1978). Differences in the low frequency Love numbers resulting from different viscosity models, turned out not to be essential. The following reflections clarify this result. The shear relaxation constant for a Maxwell rheology is $\tau=\mu/\eta_\mu$ (μ: shear modulus, η_μ: shear viscosity) (Jaeger, 1978). μ is of the order of 10^{11} Pa, η_μ is of the order of 10^{22} P. Hence, the shear relaxation constant is of the order 10^{11} s. This is small compared to the time constant involved in the Earth's slowly varying rotation.

The frequency dependent Love numbers are shown in figure 1 in the case of constant bulk modulus. Figure 1a shows the real part and figure 1b the absolute imaginary part of the Love numbers as a function of period T ($T=2\pi/\omega$). At a period of approximately 30 hours the Love numbers are extraordinarily high. The Love number h is about 5, which is not indicated in the graph. These extreme high numbers are interpreted as resonance effects connected to the top layer in model PREM being water (Denis, 1977). These periods are, however, not important for our calculations. The imaginary parts of both Love numbers are almost identical and vanish at low and high frequencies.

Accurate quantitative results for the Earth's relaxation in bulk are not available, although several authors (e.g., Herzfeld and Litovitz, 1959; Anderson, 1980; Heinz et al., 1982) have presented clear evidence for the existence of this kind of relaxation. Thus, at present, it seems impossible to consider a reasonable rheology for relaxation in bulk. Hence, only the constant bulk modulus approximation will be used here.

In the fluid case the shear moduli are zero, yielding the frequency independent hydrostatic Love numbers:

$$h_2 = 1.9333 \pm 0.0005$$
$$k_2 = 0.9333 \pm 0.0005$$

As in the viscoelastic case, they apply to the case of unrelaxed bulk moduli.

Displacements

The general formulas for displacements using Love numbers are given in Melchior (1978). If we allow for a viscoelastic Earth, the calculations require the use of frequency or time dependent Love numbers. The radial displacement R of the solid Earth is

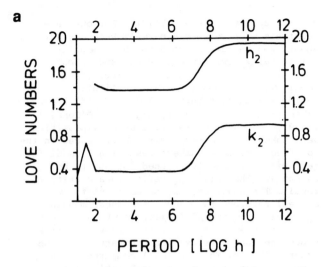

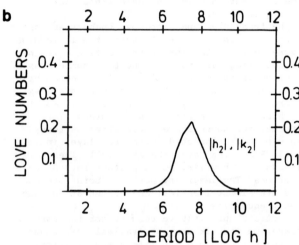

Fig. 1 Theoretical complex Love numbers in the case of unrelaxed bulk modulus as a function of period $T(T=2\pi/\omega)$. The abscissa is logarithmically scaled. The numbers indicate the hours h as a power of 10. The computations are based on the PREM Earth model. By using a Maxwell rheology for shear and the application of the correspondence principle the Love numbers were computed.
a) real part b) absolute imaginary part of the Love numbers.

$$\Delta R = \sum_{n=2}^{\infty} h_n(t,R) * \frac{U_p^n(t,R)}{g(R)} \qquad (4)$$

(U_p:primary external potential; g:gravity; h_n, k_n:Love numbers, n:degree of the potential expansion; the asterisk indicates a convolution; t:time).

The primary displacement of the Earth's potential field is

$$\xi_p = \sum_{n=2}^{\infty} \frac{U_p^n(t,R)}{g(R)} \qquad (5)$$

A secondary potential, which can be described by means of Love number k, is induced by this primary displacement.

The total displacement of the Earth's potential is

$$\Delta\xi = \sum_{n=2}^{\infty} (1+k_n(t,R)) * \frac{U_p^n(t,R)}{g(R)} \qquad (6)$$

The potential's expansion in terms of order 2 (n=2) is the dominant one, if the external potential is the centrifugal potential. Hence, in the following computations we will neglect other terms. The external potential U_p is described by the difference of the potential of the undeformed Earth $V(R)$ and the deformed Earth $V(R')$. $V(R')$ can be expanded in terms of $V(R)$ and the displacement ξ_p

$$\begin{aligned} U_p &= V(R) - V(R') \\ &= V(R) - \left[V(R) + \frac{\partial V}{\partial r}\bigg|_R \xi_p + \frac{1}{2}\frac{\partial^2 V}{\partial r^2}\bigg|_R \xi_p^2 \right] \end{aligned} \qquad (7)$$

After solving the equation for ξ and applying Dirichlet's theorem (Lambeck, 1980), the total displacement of the equipotential surface becomes

$$\Delta\xi = \frac{-\frac{\partial V}{\partial r}\bigg|_R - \sqrt{\frac{\partial V}{\partial r}\bigg|_R^2 - 2\frac{\partial^2 V}{\partial r^2}\bigg|_R \left[1+\left(\frac{R}{r}\right)^5 k\right] * U_p(r)}}{\frac{\partial^2 V}{\partial r^2}\bigg|_R} \qquad (8)$$

A plus sign in front of the root is also allowed mathematically, but this is physically not reasonable. As Dirichlet's theorem can only be utilized to describe the potential field it is not applicable to the computations of the Earth's solid surface displacements.

Centrifugal Force

The main reason for the deviation of the Earth's figure from spherical shape is the centrifugal force of the Earth's rotation. This force can be derived from the centrifugal potential U_c, which is the sum of a radial U_r and a harmonic part U_h

$$\begin{aligned} U_c &= 1/2 \Omega^2 r^2 \sin^2\theta \\ &= 1/3 \Omega^2 r^2 - 1/3 \Omega^2 r^2 P_2(\cos\theta) \\ &= \quad U_R \quad + \quad U_h \end{aligned} \qquad (9)$$

P_2:Legendre polynomial of degree 2; $\Omega = \Omega(t) = \Omega_0 + \dot{\Omega} t$:angular velocity; $\Omega_0 = 7.292115 \cdot 10^{-5}$ rad/s (Moritz, 1980); $\dot{\Omega} = -(5.4 \pm 0.5) \cdot 10^{-22}$ rad/s² (Bursa, 1984); r:radius; θ:colatitude).

The radial part U_r causes a radial displacement ξ_r of the equipotential surface which results in a change of volume (Rochester and Smylie, 1974). The harmonic part U_h causes a primary harmonic displacement ξ_p and it also induces a secondary displacement ξ_s. These displacements are not connected to a change in the volume. Furthermore, the harmonic part causes a displacement of the solid Earth ΔR. Figure 2 summarizes these deformations.

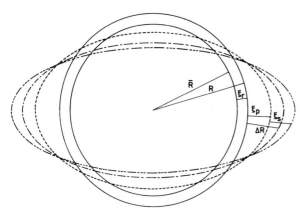

Fig. 2 The deformations of the Earth with the constant radius $\bar{R}$. The points P_0, P_1, P_2, and P_4 are on a surface of the same potential, and P_3 is a point of the solid surface of the Earth, which has not necessarily the same potential.
R: mean radius of the deformed Earth
ξ_r: purely radial displacement of the potential field
ξ_p: primary harmonic displacement of the potential field
ξ_s: secondary harmonic displacement of the potential field
ΔR: displacement of the solid Earth.

Theoretical Derivation of the Earth's Shape

The points P_0, P_1, P_2 and P_4 in figure 2 have the same potential, whereas P_3 is a point of the solid surface of the Earth whose potential is not necessarily identical to the others. Consequently, the following relation hold true:

$$V(P_0) = \frac{GM}{\bar{R}} \quad (10)$$

$$= V(P_1) + U_R(P_1) = \frac{GM}{R} + 1/3\, \Omega^2 R^2 \quad (11)$$

$$= V(P_4) + U_R(P_4) + U_h^P(P_4) + U_h^s(P_4)$$

$$= \frac{GM}{r} + 1/3\,\Omega^2 r^2 + \left[1+\left(\frac{R}{r}\right)^5 k_2\right] * \{-1/3\,\Omega^2 r^2 P_2(\cos\theta)\} \quad (12)$$

G: gravity factor, M: Earth's mass, V: potential of the non-rotating Earth, U: additional potentials due to rotation.
Knowing the mean radius R of the present Earth, i.e. 6371000.8m (according to GRS80), equations (10) and (11) yield the constant reference radius R. The use of equations (10) and (11) again yield the time dependent mean radius R(t), as well as the purely radial deformation: 6371000.8 m (according to GRS80), equations (10) and (11) yield the constant reference radius $\bar{R}$. The use of equations (10) and (11) again yield the time dependent mean radius R(t), as well as the purely radial deformation:

$$\xi_r(t) = R(t) - \bar{R} \quad (13)$$

Finally, by using equation (13) for the computation of ξ_r and by using equation (8) for the computation of $\Delta\xi$
($U_p = -1/3\, \Omega^2 \, r^2 \, P_2(\cos\theta)$) we derive

$$r(t,\theta) = \bar{R} + \xi_r(t) + \Delta\xi(r,t,\theta) \quad (14)$$

for the radii of the equipotential surface as a function of time and colatitude.
(N.B. For reasons of clarity equation (14) is not solved for r). Equation (14) holds true for the shape of the equipotential surface; hence, it is applicable for the Geoid and Hydroid.
The first order derivation of the time and colatitude dependent radii of the equipotential surface is similar to equation (14), but it is necessary to use equation (6) for the computation of $\Delta\xi$. The radii of the solid Earth are given by using equation (4)

$$r(t,\theta) = \bar{R} + \xi_r(t) + \Delta R(r,t,\theta) \quad (15)$$

Only in the case of equilibrium does the relation 1+k=h hold true. In this case we have to distinguish between the unrelaxed Hydroid (fluid Earth $\mu=0$, unrelaxed bulk modulus $K=K_o$) and the relaxed Hydroid (fluid Earth, relaxed, but not fixed bulk modulus $K=K_r$). In the case of non-equilibrium (1+k) does not equal h. Consequently, we must distinguish between the figure of an equipotential surface which is described by the Love number k and yields the Geoid, and the figure of the physical Earth which is described by the Love number h and yields the Topoid. The Geoid is the figure of that equipotential surface which best approximates the mean sea surface (extrapolated under the continents). The Topoid is the figure of the Earth's mean physical surface including oceans and continents. Therefore, the Topoid is not connected to an equipotential surface. An overview of these figures is given in the flow chart.

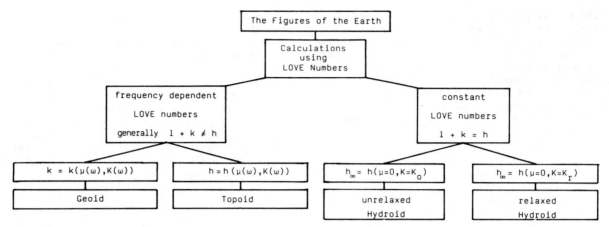

Flow chart Overview of the differentiation of the various figures. In the hydrostatic case the relation 1+k=h which is shown on the rigth side holds true. Depending on whether or not the bulk modulus is relaxed (K=K_r) or unrelaxed (K=K_0) we can differentiate between the relaxed Hydroid and unrelaxed Hydroid. The left side shows the case of frequency dependent bulk and shear moduli which result in frequency dependent Love numbers.

The flattening is given by equation

$$f(t) = \frac{r(t,90°) - r(t,0°)}{r(t,0°)} \qquad (16)$$

The computation of the Hydroid is straightforward because the Love number is independent of time. The computation of the Geoid demands the consideration of time dependent Love numbers. This results in an inhomogeneous non-linear system of equations. A linearization of that system results in an error of the same magnitude as the improvement in precision by using the second order expansion. We, therefore, compute the shape of the Geoid in first order.

However, in order to compare the Geoid and the Topoid both figures have to be computed in the same order to obtain consistent results.

Taking the history of the Earth's rotation into account and by assuming a linear variation of its angular velocity (Bursa, 1984), we can compute the displacements. After a Fourier Transformation the convolution in equation (4) and (6) are solved in the frequency domain. An inverse Fourier Transformation yields the time dependent displacement. These transformations cause problems with the initial condition and the Love number for zero frequency, but we were able to show that, in our case, these influences were negligible.

Results for the Geoid, Topoid, and Unrelaxed Hydroid

By applying the theory outlined in the previous chapters we were able to reach the following results. For the unrelaxed Hydroid (fluid case, unrelaxed bulk modulus) we obtained computations to second order for the flattening (equation 16) and the radii (equation 14) by using the Love number k=0.9333

$$a=(6378104\pm2.2)\text{m}$$
$$c=(6356823\pm2.2)\text{m}$$
$$1/f=299.71\pm0.16$$

Equation (12) and the expansion of the potential in terms of J_2 yield J_2 as a function of the Love number k (compare Lambeck, 1980). The result for J_2 of the unrelaxed Hydroid is:

$$J_2=(1.07084\pm0.001)\ 10^{-3}$$

These results are in excellent harmony with those of independent computations by the other authors whose results have been shown in table 1. Particularly those computations by Denis (1986) should be mentioned because he used the same Earth model in his third order computation as we have utilized here.

Denis computed a value of 299.68 for the inverse hydrostatic flattening. The difference is only about 0.01%.

In order to compare the Hydroid with the Topoid it was necessary to compute the Hydroid to first order. The results are:

$$a=6378099\text{ m}$$
$$c=6356804\text{ m}$$
$$1/f=299.52$$

Since the Earth's angular velocity and, consequently, the centrigual forces are functions of time the Hydroids are time dependent. Taking the time dependency into account equation (14) yields the radii of the Hydroids as a function of time. The results of these computations are shown in figures 3a-c. The uncertainty of the angular acceleration is indicated by $\dot{\Omega}+10\%$ and $\dot{\Omega}-10\%$. As expected the flattening decreases. Furthermore, it should be noted that the volume of the hydrostatic figure also decreases. In the case of zero angular velocity the volume will be minimal, i.e. $V=4/3\pi\bar{R}^3$.

By utilizing frequency dependent Love numbers the Topoid and Geoid were computed as a function of time by means of Fourier Transformations. The different investigated shear viscosity models yielded similar results which is reasonable if we see the argumentation in the section about Love numbers. The first order results for the theoretically present Geoid are:

$$a = (6378099 \pm 2.2) \text{ m}$$
$$1/f = 299.51 \pm 0.16$$
$$J_2 = (1.07087 \pm 0.001)\, 10^{-3}$$

and for the theoretically present Topoid

$$a = (6378098 \pm 2.2) \text{ m}$$
$$1/f = 299.56$$

A comparison of those results with those of the first order unrelaxed Hydroid shows no significant difference. Proceeding on the suppositions outlined above all three figures are identical. Thus, we conclude that on the premises of relaxation in shear alone and neglecting internal dynamic forces no difference exists between the flattenings of the theoretical Topoid, Geoid, and Hydroid. The shear moduli are absolutely relaxed, and all three figures reflect hydrostatic equilibrium. This result clearly shows that the Earth is not able to support a fossil equatorial bulge.

The time dependence of the Geoid and Topoid is similar to those in figures 3a-c for the Hydroids. Minor differences, however, are so small that they cannot be illustrated in the graphs.

Taking the error limits into account the rate of changes are approximate constants, and the following results were obtained:

$$\dot{R} = -(3.4 \pm 0.3)\, 10^{-6} \text{ m/yr}$$
$$\dot{a} = -(6.8 \pm 0.7)\, 10^{-6} \text{ m/yr}$$
$$\dot{V} = -(1.8 \pm 0.2)\, 10^{9} \text{ m}^3/\text{yr}$$
$$\dot{J_2} = -(5.0 \pm 0.5)\, 10^{-13}\, 1/\text{yr}$$
$$\dot{1/f} = +(1.4 \pm 0.1)\, 10^{-7}\, 1/\text{yr}$$

However, if we compare the results with those from measurements of the present Geoid - according to the GRS80:

$$a = 6378137 \text{ m}$$
$$1/f = 298.257$$
$$J_2 = 1.08263\, 10^{-3}$$

- a clear difference becomes evident. As shown above, this difference cannot be attributed to

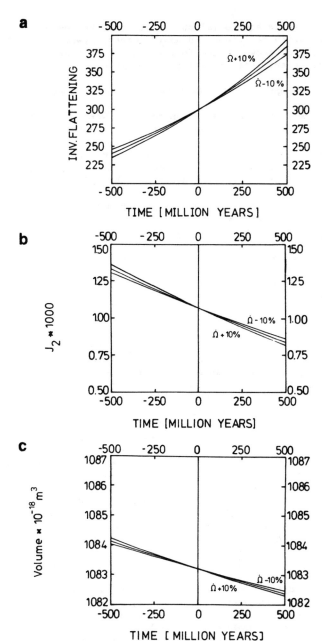

Fig. 3 The time dependent hydrostatic figure with unrelaxed bulk modulus. The computations were carried out to second order and are based on a linear variation of the Earth's rotation. $\dot{\Omega}+10\%$ and $\dot{\Omega}-10\%$ indicate the uncertainty of the angular acceleration.
Time=0 indicates the present.
a) flattening as a function of time
$f(t=0)=1/299.71$
b) dynamical form factor as a function of time
$J_2(t=0)=0.00107084$
c) volume as a function of time
$V(t=0)=10821\, 10^{21} \text{ m}^3$

a fossile equatorial bulge. It could be the consequence of internal dynamic forces. However, it will be shown in the following section that a different interpretation is possible.

Ambiguity of the Earth's Hydrostatic Figure

All the results presented above were obtained on the assumption that there is no relaxation in bulk. However, it was mentioned already that there is indication that this assumption may not be justified especially when long term deformations of the Earth are considered. As the bulk modulus structure of the Earth influences both Love numbers h and k, one would expect the figure of the Earth to be dependent on the degree of relaxation in bulk. This is true for all three figures, the Hydroid, the Geoid and the Topoid. For the practical computation of these figures it would mean that the seismic bulk modulus could no longer be used to determine the Earth's response to centrifugal forces but a long term or static bulk modulus would have to be applied. Also, the density structure would have to be modified because it is closely related to the bulk modulus by the Adams-Williamson equation (Stacey, 1977). This equation incorporates the seismic parameter $\Phi = K/\rho$ (K:bulk modulus), as a basis for determining the density structure. It is commonly derived from seismic measurements Hence, the seismic (unrelaxed) bulk modulus enters into the derivation of the density structure and not the static one as it should be in the case of slow long term deformations of the Earth. Thus, all computations of the Earth's hydrostatic figure may be wrong, because they are based on an unrelaxed density structure, which is not corrected for the effect of bulk relaxation. As a consequence, the deviation of the observed Geoid from the unrelaxed hydrostatic figure cannot necessarily be taken as an indication for the importance of dynamic forces within the Earth's interior. It could simply be an effect of relaxation in bulk. The implementation of relaxation in bulk would result in a different Hydroid which could have the same shape as the Geoid. Thus, we have to look for a new criterion which allows to decide whether the Earth is in hydrostatic equilibrium or not. Such a criterion will be proposed in the next section.

An Unequivocal Criterion for the Hydrostatic Equilibrium

Quantitative results for the Earth's relaxation in bulk are not available. Thus, at present it seems impossible to theoretically compute the figures by considering the relaxation in bulk. However, another criterion for the determination of hydrostatic equilibrium can be introduced. This criterion is based on the comparison of the Topoid and the Geoid, figures which can be directly measured. If both figures are not identical the vector of gravity would be normal to the Geoid, but not normal to the surface of the solid earth. This would result in internal stresses, and the Earth could not be in equilibrium. Only if the Topoid and the Geoid were the same, the Earth would be in hydrostatic equilibrium.

Satellite data are avialable for the Geoid (Moritz, 1980), but not for the Topoid. Using topographical measurements it is possible in practice to expand the Topoid in spherical harmonics. The measurements of the Topoid should, however, be independent of a reference figure which is based on the Earth's potential field. Some authors have made an analysis of the Earth's topography in spherical harmonics (Balmino et al., 1973; Seidler, pers. comm., 1987) but these expansions cannot be utilized for the comparison of the Topoid and Geoid because neither the degree 2 expansion of the Topoid dominates nor does the expansion converge rapidly enough. Hence, the question of the Earth's hydrostatic equilibrium cannot be answered even with the new criterion unless better information about the Topoid is available.

Discussion and Conclusions

Considering the Earth's response to centrifugal forces, a new method for the computation of the Earth's figures has been introduced. This theory is based on the rheological properties of the Earth and allows the computation of the Geoid, Topoid, and the Hydroid with the help of frequency dependent Love numbers. The comparison of these figures suggests that the Earth is in hydrostatic equilibrium as far as its response to the centrifugal forces is concerned. However, in comparing the observed Geoid with the shape of the unrelaxed Hydroid, a clear difference becomes evident.

Attempts have often been made to determine the minimum strength of the Earth by investigating the stress differences which would result from a deviation of the hydrostatic figure (computed with the classic theory) and the Geoid (e.g. Jeffreys, 1943; Jeffreys, 1963; Zharkov and Trubisyn, 1978; Nakiboglu, 1982). However, we do not believe that this difference is necessarily an indication of the Earth's deviation from hydrostatic equilibrium. It could as well be a consequence of relaxation in bulk. This kind of relaxation would imply a dynamic density structure. The density structure as derived from seismic velocities would not be applied to the Earth's response to long term external forces, but a static (relaxed) one would be relevant. As the degree of relaxation in bulk is not yet known, we were not able to compute the relaxed Hydroid (relaxed refers here to relaxation in bulk). Only the new criterion, the comaprison of the Geoid and Topoid, is able to give evidence concerning the Earth's equilibrium. It is precisely the comparison of Topoid and Geoid. Only when these figures are identical can equilibrium exist, otherwise the vector of gravity which is always normal to the surface of the Geoid would not be normal to the surface of the Topoid.

Acknowledgements. For comments and critical review of the manuscript we would like to thank J. Voss. We are grateful to J. Welling, L. Bittner, and K. Helbig for their help during the preparation of the manuscript. (Publication No. 363, Institute of Geophysics, Kiel).

References

Anderson, D.L., Bulk attenuation in the Earth and the viscosity of the core, Nature, 285, 204-207, 1980.

Balmino, G., K. Lambeck, and W.M. Kaula, A spherical harmonic analysis of the Earth's topography, J. Geophys. Res., 78, 478-481, 1973.

Budo, A., Theoretische Mechanik, VEB Deutscher Verlag der Wissenschaften, 9. Auflage, 1978.

Bursa, M., Secular Love numbers and hydrostatic equilibrium of planets, Earth Moon and Planets, 31, 135-140, 1984.

Cathles, L.M., The viscosity of the Earth's mantle, Princeton Univ. Press, Princeton, 1975.

Clairaut, A.C., Théorie de la figure de la Terre, Tirée de principes de l'hydrostatique, Courcier, Paris, 1743.

Denis, C., Static and dynamic effects in theoretical Love numbers, in: Proceedings of the 8th. int. symposium on Earth tides, published by M.Bonatz and P.Melchior, 709-729, 1977.

Denis, C., The computation of equilibrium figures and the hydrostatic flattening of the Earth, Proc. J.G.L., 48, 16-19, 1981.

Denis, C., The hydrostatic figure of the Earth and the reference Earth model, Proc. J.G.L., 52, 1-10, 1983.

Denis, C., The hydrostatic figure of the Earth, in: Physics and Evolution of the Earth's Interior, Vol. 4, edited by R.Teisseyre, publ. jointly by PWN, Warsaw and Elsevier, 1986.

Dziewonski, A.M., and D.L. Anderson, Preliminary reference Earth model, Phys. Earth Planet. Interiors, 25, 297-356, 1981.

Hager, C.L., and R.J. O'Connell, Kinematic model of large scale flow in the Earth's mantle, J. Geophys. Res., 84, 1031-1048, 1979.

Heinz, D.R., Jeanloz, and R.J. O'Connell, Bulk attenuation in a polycristaline Earth, J. Geophys. Res., 87, 7772-7778, 1982.

Herzfeld, K.F., and T.A. Litovitz, Absorption and dispersion of ultrasonic waves, Academic, New York, 1959.

Jaeger, J.C., Elasticity, Fracture and Flow, with engineering and geological application, Science Paperback, London, 1978.

James, R., and Z. Kopal, The equilibrium figures of the Earth and the major planets, Icarus, 1, 442-454, 1963.

Jeffreys, H., The stress differences in the Earth's shell, Mon. Not. Roy. astr. Soc. Geophys. Suppl., 5, 71-89, 1943.

Jeffreys, H., On the hydrostatic theory of the figure of the Earth, Geophys. J. Roy. astr. Soc., 8, 196-202, 1963.

Khan, M.A., General solution of the problem of hydrostatic equilibrium of the Earth, Geophys. J. Roy. astr. Soc., 18, 177-188, 1969.

Lambeck, K., The Earth's variable rotation: geophysical causes and consequences, Cambridge Universtiy Press, Cambridge, 1980.

Lanzano, P., Deformations of the elastic Earth, Academic Press, 1982.

Laplace, P.S., Mécanique Céleste, 5, Paris, 1825.

Love, A.E.H., The yielding of the Earth to disturbing forces, Proc. of Roy. Soc., series A, 82, 73-88, 1909.

Moritz, H., Geodetic Reference System 1980, Bull. Geod., 54, 395-405, 1980.

Nakiboglu, S.M., Hydrostatic figures and related properties of the Earth, Geophys. J. Roy. astr. Soc., 57, 639-648, 1979.

Nakiboglu, S.M., Hydrostatic theory of the Earth and its mechanical implications, Phy. Earth Plan. Int., 28, 302-311, 1982.

Peltier, W.R., The impulse response of a Maxwell Earth, Rev. Geophys. Space Phys., 12, 649-669, 1974.

Rochester, M.G., and D.E. Smylie, On change in the trace of the Earth's inertia tensor, J. Geophys. Res., 79, 4948-4951, 1974.

Stacey, F.D., Physics of the Earth, 2nd edition, John Wiley & Sons, New York, 1977.

Tonn, R., On the figures of the Earth, Proc. of the tenth int. symp. on Earth tides, 415-422, 1974.

Vetter, U., Stress and viscosity in the astenosphere, Tectonophysics, 56, 145-146, 1979.

Wang, R., Das viscoelastische Verhalten der Erde auf langfristige Gezeitenterme, Diplomarbeit, Universität Kiel, 1986.

Zschau, J., Auflastgezeiten, Habilitationsschrift, Universität Kiel, 1979.

CONTEMPORARY VERTICAL CRUSTAL MOTION IN THE PACIFIC NORTHWEST

Sandford R. Holdahl[1], Francois Faucher[2], and Herb Dragert[3]

Abstract. A map of recent vertical crustal motion has been compiled for coastal Washington and southwest British Columbia. Average velocities over the past 80 years were determined by least squares adjustment of repeated precise levelings and mean sea-level observations from 21 tide gauges. Annual variations in mean sea level were determined directly within the adjustment model under the assumption that they were identical at all tide gauges in a given year. The derived vertical velocities range from -2.0 ±0.9 mm/yr near Seattle, to 2.5 ±0.8 mm/yr at the northwest tip of the Olympic Peninsula, and over 3 mm/yr in the region to the north of Campbell River on Vancouver Island. Determination of a constant velocity in this latter region is complicated by about 10 cm of coseismic subsidence associated with a magnitude 7.3 earthquake in 1946, and an apparent increase in uplift rate over the past decade. Qualitatively, the regional velocity pattern is consistent with features of the current plate convergence model: 1) the rapid uplift of the region north of central Vancouver Island is consistent with the overriding of the young (<6My), buoyant Explorer Plate which may be underplating the coastal margin in this area; and 2) the ridge of uplift extending from the Neah Bay area north across Vancouver Island to Campbell River is consistent with a pattern expected from a locked subduction zone underlying this coastal region. The large-scale subsidence to the southwest of Puget Sound is more difficult to explain in the context of plate convergence.

Introduction

Secular vertical motion of a point on the earth's surface can be defined as the component of vertical motion at that point which is unchanging over geological periods of time. As such, it is a mean velocity over thousands of years. To estimate this velocity by analysis of geodetic measurements carried out over time intervals of a few decades is extremely difficult in active seismic areas. In such areas, vertical motion will be determined largely by the current phase of the deformation cycle associated with the repeated occurrence of large crustal earthquakes (Thatcher and Rundle, 1984) as well as possible glacio-isostatic readjustment. In the Pacific Northwest the cycle time for smaller but still deformation-producing events (M~7) is roughly 30 to 80 years (Milne et al., 1978), with three such earthquakes having occurred on central Vancouver Island this century (1918, 1946, and 1957). The return period for possible megathrust (M>8) earthquakes in this region is estimated at hundreds of years (Rogers, 1988). Since the data of the present study do not span a complete earthquake cycle and since post-glacial readjustment in this region is virtually complete (Mathews et al., 1970), it is likely that our derived vertical velocities are dominated by "interseismic" strain rates; i.e., elastic and visco-elastic behaviour of the crust in the time interval between major earthquakes. Consequently, the "snapshot" of contemporary vertical crustal motion provided by geodetic data is invaluable in assessing the current dynamic state of local plate interactions. Furthermore, the determination of the regional

---

[1] National Geodetic Survey, Charting and Geodetic Services, National Ocean Service, NOAA, Rockville, Maryland 20852
[2] Canada Centre for Mapping; Surveys, Mapping and Remote Sensing Sector; Energy, Mines, and Resources, Ottawa, Canada K1A 0E9
[3] Geological Survey of Canada, Pacific Geoscience Centre, P.O. Box 6000, Sidney, B.C. V8L 4B2

Copyright 1989 by
International Union of Geodesy and Geophysics and American Geophysical Union.

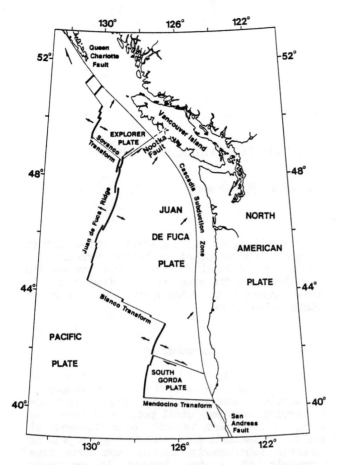

Fig. 1. Plate tectonic setting of the Pacific Northwest. Of particular significance to vertical deformation studies in the Cascadia Subduction Zone is the change in the orientation of the coastline as one moves north, the presence of 3 small distinct young oceanic plates, and the question of whether the subduction zone is currently locked along its entire length.

pattern of velocities may help to evaluate and refine descriptions of the tectonic setting in the Pacific Northwest (see Figure 1).

Tide gauge data from southwest British Columbia have previously been used to determine vertical motion (De Jong and Siebenhuener, 1972; Wigen and Stephenson, 1980; Riddihough, 1982). Vanicek and Nagy (1981) used mean sea-level data and releveling data to compile a map of vertical crustal movements across Canada. The present study improves the part of that map covering the Pacific northwest by utilizing new leveling data and longer tidal records. The recent leveling data on Vancouver Island have been discussed in detail by Faucher and Blackie (1986), and elevation changes observed in central Vancouver Island have been analyzed by Dragert (1986, 1987). Of particular concern in these latest studies has been the correction of leveling data for systematic effects due to rod errors, refraction errors, and magnetic errors. Leveling and tide gauge data in the coastal Washington region has been discussed by Balazs and Holdahl (1974), Ando and Balazs (1979), Holdahl and Hardy (1979), Reilinger and Adams (1982), and Holdahl et al. (1986).

The present investigation combines the Canadian and U.S. data from these previous studies and adds new leveling data east of Puget Sound and north of Vancouver, B.C. Mean sea-level data up to 1985 have been incorporated at most principal tide gauge sites, and subordinate tide gauge data from ten stations have also been added to improve the distribution of velocity information in the coastal Washington area. Furthermore, by including the annual variation of mean sea level (MSL) as a parameter in the adjustment model, the variances of the estimates of linear trends are significantly reduced. Refined refraction corrections have been applied retroactively to improve the accuracy of old levelings (Holdahl, 1981, 1983), and an empirical approach was taken to remove significant magnetic error on an instrument-by-instrument basis (Faucher and Blackie, 1986; Holdahl et al., 1987). Finally, coseismic deformation associated with the 1946 earthquake near Campbell River was modeled using dislocation theory (and fault parameters constrained by seismic data) in order to remove the step-function effect in vertical displacements caused by this earthquake. No allowances have been made for possible significant visco-elastic effects following the 1918 and 1946 earthquakes, and a constant vertical velocity is assumed over the time period spanned by the MSL and leveling data.

Mathematical Model

In our model for adjustment of a deforming level network, the height, H_i, at time t_i, of a geodetic station is given by

$$H_i = H_o + \dot{H}(t_i - t_o) + U + s_i \qquad (1)$$

H_o is the height of the station at t_o, a selected reference time, and U is elevation change caused by coseismic slippage on one or more fault planes (Holdahl, 1986). The s_i term accounts for the variation of sea level when a height observation is based upon a mean sea

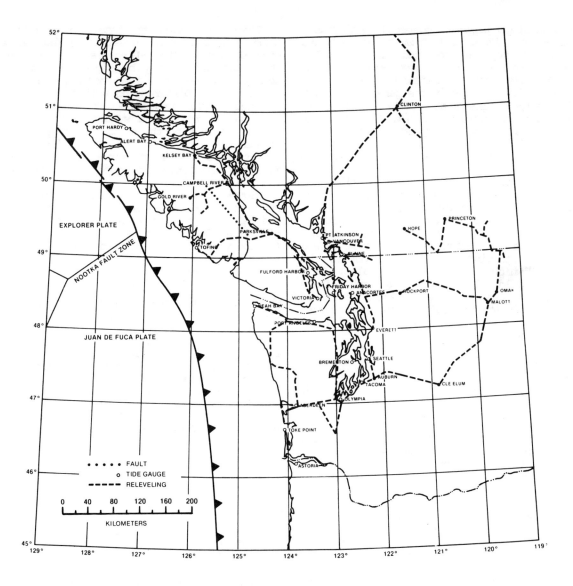

Fig. 2. Locations in the Pacific Northwest where repeated measurements have been made at tide gauges and along level lines.

level measurement (monthly or annual) but is not present in observation equations for level measurements. The change in height during the period (t_i-t_o) is considered to have occurred uniformly at the velocity $\dot{H}$. $\dot{H}$ may be mathematically expressed in any number of ways. $\dot{H}$ at a position (x,y) is calculated here using a method called multiquadric (MQ) analysis (Holdahl and Hardy, 1979):

$$\dot{H} = \Sigma r_j [(x-x_j)^2 + (y-y_j)^2]^{1/2} \qquad (2)$$

The x_j, y_j are coordinates of MQ nodal points and the r_j values are unknown coefficients. The locations of MQ nodal points are determined by review of leveling profiles to find maxima and minima of the relative motion for each route of releveling. The number of nodal points should be sufficient to portray the features seen in the profiles.

MSL Data

Our computation of heights and velocities was strengthened by incorporating data obtained at 11 control gauges which operated continuously at permanent sites, and from 10 subordinate gauges which have been operated intermittently. The locations of these gauges are shown in Figure 2. Sea level measurements are referred to a set of bench marks in the general vicinity of the gauge. The most stable of these is designated as the primary bench mark (PBM). The PBM must be stable with respect to nearby bench marks, but is allowed to move absolutely in a way that is considered representative of the region near it.

There are several methods for including sea level measurements in crustal motion computations. The absolute velocity, $\dot{H}$, at a tide gauge can be computed by deriving the trend of sea level change with respect to the PBM, correcting for the eustatic (global) rise of sea level, and changing the sign to obtain the velocity at which the land rises or falls. This procedure could also be used at subordinate gauges that are occupied for a short period of time or sporadically, but is not recommended because unknown annual variations of sea level would obscure the result.

An estimate of the relative velocity

$$\Delta \dot{H}_{1-2} = \dot{H}_2 - \dot{H}_1 \qquad (3)$$

between two control gauge sites (stations 1 and 2) can be obtained by fitting a straight line through a plot of the differences of mean sea levels. For each year, the difference is calculated as

$$\Delta W_{1-2,i} = W_{2,i} - W_{1,i} \qquad (4)$$

where $W_{1,i}$ and $W_{2,i}$ are the mean sea levels for the two stations corresponding to mean time t_i. The eustatic rise of sea level cancels, along with the annual variations of sea level that are common to both stations. The variability of the yearly differences are usually much smaller if the two gauge sites are in the same hydrodynamic regime, otherwise little or no cancellation of annual variations will occur. If the MSL values from a station are differenced with values from more than one other station, the computed relative velocities are correlated, and a non-diagonal weight matrix is required to incorporate them as observations in a level network adjustment. Although it is sometimes most convenient to use absolute velocities, Wigen and Stephenson (1980) point out that differencing of MSL values from two gauge sites will give a more precise determination of relative velocities through common mode noise rejection.

Subordinate tide stations are occupied for short periods of time, generally a few months to several years. The MSL values observed during these occupations may be higher or lower than normal because of the monthly or annual variability. Differencing with a control gauge is the usual procedure for computing the relative velocity to a subordinate gauge. But in the Pacific northwest where several control gauges are available for comparison, it is advantageous to use yet another approach wherein MSL data are converted to observed heights of the PBM.

Figure 3 shows the relationship between the PBM and the observed mean water level. H_d is the height of the PBM with respect to the "zero" of the staff. The observed height of the PBM for the year t_i can be written as

$$H_i = H_d - (W_i - m_o(t_i - t_o) - s_i) + h^* \qquad (5)$$

where h^* is the sea surface topography, i.e., the permanent elevation of mean water level above the reference equipotential surface corresponding to zero height. m_o is the eustatic rise of sea level. s_i is the annual variation of sea level which is common to all control and subordinate tide gauges in the same region. We do not yet know the s_i values, but can include them in our adjustment model (1). The expression for an observed height of the PBM, derived from a mean sea level observation, is

$$H_d - W_i + m_o(t_i - t_o) + h^*$$
$$= H_o + \dot{H}(t_i - t_o) + U - s_i \qquad (6)$$

The left-hand side consists of measured or assumed values. In this adjustment model, zero height corresponds to adjusted MSL at t_o. The weight for this type of observed height should be approximately $(2 \text{ cm})^{-2}$, knowing that the annual variation will be mostly removed.

The s_i are not uniquely determined. We propose the particular solution which has the property

$$\Sigma s_i^2 = \text{minimum}.$$

This is accomplished by adding the following weighted constraints (quasi-observations):

$$s_i = 0, \quad i = 1, k$$

where k is the number of years for which mean sea level data are available. The quasi-observations of s_i are given a low weight of $(15 \text{ cm})^{-2}$ to minimize their influence in the adjustment.

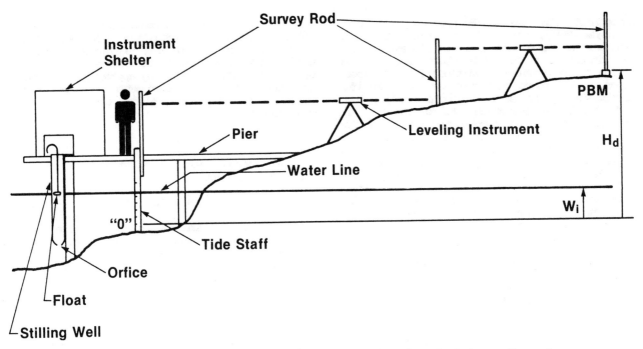

Fig. 3. Typical tide gauge installation showing relationship between the primary bench mark (PBM), mean water level, and the zero of the tide staff.

The advantages of this method are
1. the annual variations are derived from all possible control tide gauges in the same hydrodynamic regime;
2. errors are propagated rigorously without resorting to non-diagonal weight matrices, and all data are fully utilized;
3. leveling can help to compute velocities and hence annual variations; and
4. the bias in MSL values caused by coseismic motion can be removed by the adjustment model.

By modeling the annual variations in this way we do not address the causes of annual sea level variations (see Vanicek, 1978), we merely wish to remove their influence because they are a nuisance. Additionally, our assumption of no sea surface topography should not be construed as disbelief in its existence, but rather recognition that its variation is small over short distances and leveling is too weak to define its magnitude over long distances. A good discussion of the variables, data requirements, and computational techniques involved in estimating h* may be found in Merry and Vanicek (1982).

Minimization of Systematic Leveling Errors

Leveling has always been one of the most precise geodetic measurements. The standard deviation of a levelled height difference is expressed as $aS^{1/2}$, where S is equal to the distance levelled in kilometers. The value "a" is the standard deviation for levelling of 1 kilometer, and, since 1900, has been estimated to be approximately 1 ±.5 mm. The change in height difference between two monumented points 200 km apart is obtained by leveling that distance at two different times. The standard deviation of the measured change in relative height between the two points would be approximately 2 cm. Unfortunately, this precision is seldom achieved because systematic error becomes a dominant contribution over distances greater than about 50 km. Therefore, our primary concern in estimating crustal motions over long distances must be with systematic error.

The leveling data in the Pacific Northwest have been analyzed cautiously under the assumption that they are biased by the usual small accumulations of systematic leveling errors. However, errors which are correllated with height should have only a minimum impact on this study because so many of the levelings follow low-lying coastal routes. The levelings are adjusted with the MSL data to ensure consistency, thus growth of systematic error near the coast is controlled to a large degree.

Refraction is the largest of the height-correllated errors, and its scale is dependent primarily on the amount of solar radiation throughout the day. Because of the high latitude (49°), the angle at which the sun's rays strike the ground would be 15 degrees less than in southern California where refraction is a significant problem. Additionally, the cloud-cover and ground moisture in the Pacific Northwest help to minimize upward sensible heat flux, and hence refraction. All of the leveling data prior to 1982 were corrected for refraction errors using the U.S. and Canadian models to estimate vertical temperature stratification. These models are based on solar radiation records and are described by Holdahl (1982) and Heroux et al (1985). Because temperature measurements were not actually made, and other approximations may have been required to determine sightlengths in old leveling, the possibility of significant residual refraction error exists on the few level lines which traverse high topography. After 1982, the Canadian data were corrected for refraction using observed vertical temperature differences. The two level lines which are directed east from Puget Sound, then join and bend north into Canada, may have sufficient residual refraction error to obscure the real pattern of vertical motion. This argument is equally applicable to the long line directed north from Vancouver.

Magnetic error is the cause of most concern in this study. Many of the most recent levelings were performed between 1972 and 1978 using Zeiss Ni-1 levels which have since been found to be subject to influence by magnetic fields (Rumpf and Meurisch, 1981). The resulting systematic error may be as large as 1 or 2 mm/km when the leveling is directed towards magnetic north or south, but will be zero when directed towards magnetic east or west. The magnetic error correction constants derived in laboratory calibrations have been rarely successful. Consequently, Holdahl et al. (1987), and Faucher and Blackie (1986) have used empirical methods to calibrate the Zeiss Ni-1 levels. This study used such empirically derived calibration constants to correct for the magnetic error of Zeiss Ni-1 data.

In addition to the random and systematic error of the leveling measurement, there is also the question of whether the monuments at any two points are capable of maintaining relative stability. Review of many profiles of relative elevation change has shown that nearly all bench marks undergo some amount of local movement. This can be called "ground noise," and it is caused by such factors as frost, variations in soil moisture, vibration, ground slumping,, etc. These are nuisance motions which can somewhat cloud our determination of the regional motion pattern. The cause of ground noise at one point may be unrelated to the cause at a point one kilometer away. Ground noise tends to accumulate with time at a decaying rate following the establishment of most monuments.

Ground noise weakens our computation of vertical crustal motion in much the same way that imprecision of a leveling measurement would, therefore we can reflect the uncertainty caused by ground noise by choosing appropriate weights for the leveling measurements. The usual weight, p, for a leveling observation is expressed as

$$p = (a^2 S)^{-1} \qquad (7)$$

To account for ground noise we add a variance component

$$p = (a^2 S + b^2 (t_i - t_o))^{-1} \qquad (8)$$

where b is 2 mm. This means, for example, that in 50 years a bench mark would have a 67 percent probability of moving less than 14 mm due to very local conditions. After correcting for systematic errors as best possible, this method of weighting treats any residual systematic error and ground noise as if it is random measurement error.

Coseismic Motion on Vancouver Island

The 1946 earthquake on central Vancouver Island produced significant coseismic deformation as indicated by a comparison of the 1930 and 1946 (post-earthquake) levelings from Parksville to Campbell River (Figure 4). The coseismic motion is another nuisance as we try to determine velocities. The 1946 event produced motion opposite in direction to the post-seismic trend and thus much of it is cancelled in the total motion from 1930 to 1984. Note that this particular traverse was surveyed twice in the most recent past (1977 and 1984) and shows an apparent acceleration in relative uplift rates along this profile over the most recent survey interval (Dragert, 1987). For this study, a constant uplift rate is assumed.

The coseismic motion was modeled in a preliminary way by fitting a dislocation model to the 1930 and 1946 leveling data between Parksville and Campbell River. A complete description of this technique, as applied to leveling data, is given by Holdahl (1986). The fit was done by an automated trial and error or "jiggle" program in which all fault parameters are varied. Starting parameters were taken from Slawson and Savage (1979), which they derived from a fit to pre- and post-earthquake triangulation measurements at a small network near the epicenter. These initial values produced vertical displacements which did not fit well along the Parksville-Campbell River level line. A longer (80 km) and deeper (5-40 km) fault was needed, having less right-lateral strike slip (40 cm) and more normal slip (20 cm). Strike of this fault plane is 338°, halflength is 44 km, dip is 72.9°, and its

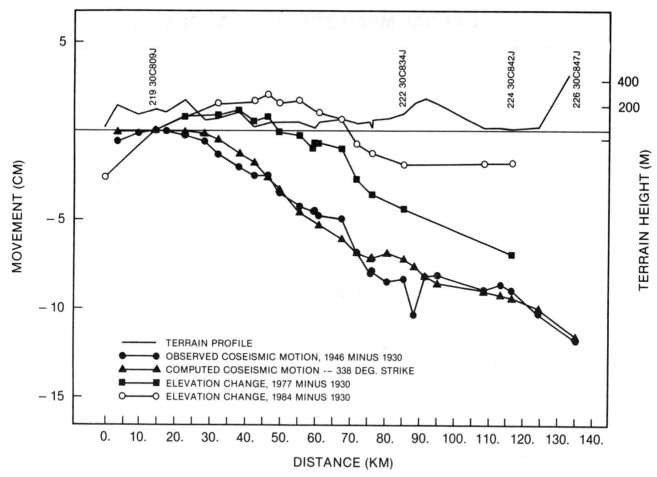

Fig. 4. Cumulative changes in elevations obtained from the four leveling surveys between Parksville and Campbell River on the east coast of central Vancouver Island. The line connecting triangles shows the computed coseismic motion obtained from the adjustment. The post-seismic uplift rate has apparently increased since 1977.

center is 49.65° latitude, 125.25°W longitude. This result agrees generally with solutions obtained previously by Rogers and Hasegawa, 1978. These fault parameters are not unique. A fault plane striking perpendicular to this plane can also produce a very good fit to the leveling data from Parksville to Campbell River. Thus our preference for this fault plane is based on arguments previously presented by Rodgers and Hasegawa. After converging to a fit for this subset of data, the fault parameters were put into the adjustment of all the leveling and tide gauge data. The values for strike and dip slip were input as observations having weights, and the other fault parameters were fixed. In the adjustment of the whole data set, the preliminary estimates of strike slip and dip slip remained virtually unchanged.

Medium-size earthquakes (Olympic earthquake of 1949, M_s = 7.1; Puget Sound earthquake of 1965, M_s = 6.5) have occurred in Washington State. These were deep earthquakes (60 km) which produced only minor and localized surface deformation, and therefore were not included in our model.

Application to the Pacific Northwest

The releveling in the Pacific Northwest is geometrically weak (see Figure 2). Only three circuits of releveling are formed, those being in Washington State. The leveling data provided 487 observed height differences. One hundred MQ nodal points were used to describe the velocity surface. It is fortunate that so many tide gauges exist in the region. The annual MSL values were used to initialize the network by providing heights and velocity information. The control and subordinate gauge data entered the adjustment as 537 observed heights of the

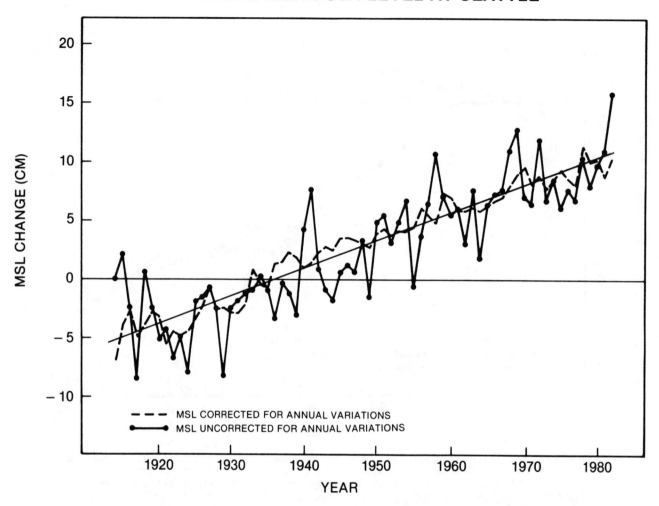

Fig. 5. Annual mean sea level at Seattle is plotted before and after removing the regional mean annual variations that were derived from analysis of re-occupied tide gauges in the Pacific Northwest. The scatter about the trend line is 67 percent less as a result of removing the mean annual variations.

tide gauge bench marks. There were two exceptions: the data from Astoria entered the adjustment as a velocity difference from Neah Bay; and Alert Bay was differenced with Port Hardy. At all gauge sites the sea surface topography, h*, was assumed to be zero, and the eustatic rise of sea level was taken to be 1 mm/yr (Nakiboglu, 1986). No error was assigned to the eustatic rise. Any error in m_0 will cause a constant bias for all computed velocities. Annual variations of sea level were resolved in the adjustment for the years from 1909 to 1983.

To test whether the annual variations of MSL were adequately removed by the adjustment process, a straight-line fit was made to corrected and uncorrected MSL data from Vancouver, Seattle, and Neah Bay. For each of these widely separated stations, the RMS error on a linear fit to the corrected data (annual variations removed) was reduced by approximately 67 percent as compared to the uncorrected data. Figure 5 is a plot of the Seattle mean sea level data in corrected and uncorrected form, which vividly shows the reduction in scatter about the trend due to removal of the estimated annual variations. We conclude that it is reasonable to assume each station is in the same hydrodynamic regime. At Neah Bay, a significant change in trend of sea level was calculated as a

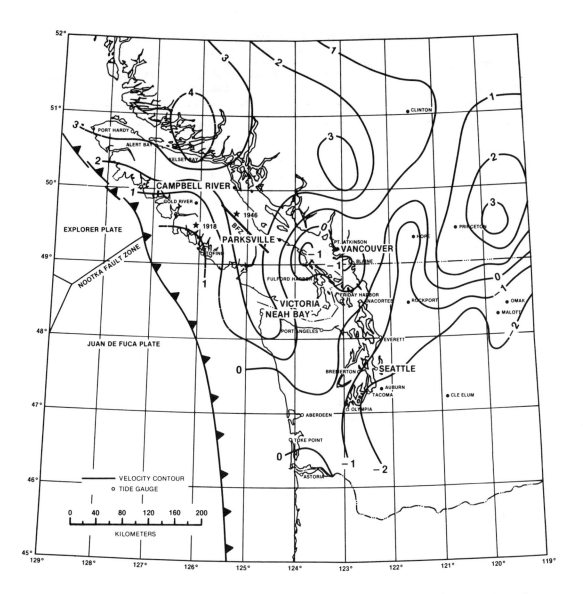

Fig. 6. Smoothed contours of vertical velocity (mm/yr) computed from the adjustment model. The epicentres of the two large on-shore earthquakes in central Vancouver Island are shown by stars annotated with the year of occurrence. The Beaufort fault zone (BFZ) is shown by a heavy line.

result of fitting to the corrected data. The trend through the uncorrected sea level data was -0.93 ±0.34 mm/yr, and changed to -1.69 ±0.11 mm/yr after accounting for the annual variations. The significant change in trend at Neah Bay was in part attributed to a strong shift in the uncorrected annual mean for 1983, which comprised the last point in a shorter data series. The computed annual mean for 1983 was 11.6 cm, and was by far the largest for any year. The trends at Seattle and Vancouver were unchanged.

Figure 6 presents a smoothed contour map of vertical velocities resulting from the adjustment, and Table 1 gives estimated point velocities and standard deviations for key locations. Before summarizing the general pattern and its possible relationship to tectonics, the question of reliability of these contours should be addressed. The velocities shown are most reliable near the coastlines where tide gauge data constrain the leveling, and where residual height-correlated errors will be a minimum. The velocity standard deviation

TABLE 1. Adjusted Heights and Velocities at Representative Locations in the Pacific Northwest

Code	City/BM Designation	Height (m)	Std.Dev. (cm)	Velocity (mm/yr)	Std.Dev. (mm/yr)	Years of MSL Observations
o	NEAH BAY, TGBM 15	4.975	2.9	2.5	0.8	`35-`83
o	FRIDAY HARBOR	3.560	2.9	0.4	0.8	`34-`83
o	PT. ATKINSON, 213J2=24C004	1.433	2.9	-0.2	0.8	`14-`84
o	SEATTLE, TGBM 7	4.597	2.9	-1.4	0.8	`14-`82
o	VANCOUVER, 1J	13.240	2.9	0.7	0.8	`10-`84
	ROCKPORT, W 60	85.199	3.2	-1.5	1.1	
	AUBURN, N	25.193	2.9	-2.6	0.9	
+	TOKE PT. TGBM 5	2.704	2.9	-1.0	0.9	`35-`39, `77-`83
+	BREMERTON, TGBM 2	3.892	3.1	-0.8	0.9	`35-`50, `77-`78
+	ABERDEEN, TGBM 5	4.376	2.9	-0.3	0.9	`34-`35`, `82-`83
+	PORT ANGELES, TGBM 11	7.250	2.9	0.4	0.9	`34-`35, `77-`83
+	EVERETT, TGBM 4	10.875	2.9	-0.9	0.8	`34-`35, `77
o	TOFINO, 78C089	2.543	2.9	1.9	0.8	`10-`84
o	VICTORIA, 29C737J	2.802	2.9	0.2	0.8	`9-`84
+	TACOMA, G 11	5.125	2.9	-2.4	0.9	`34-`35, `77-`78
+	OLYMPIA, TGBM 11	3.611	2.9	-1.0	0.9	`34-`35, `77-`78
+	BLAINE, TGBM 4	10.631	2.9	-0.4	0.9	`34-`35, `74-`75
+	ANACORTES, TGBM 6	5.658	3.2	0.2	1.0	`34-`35, `74-`75
	PARKSVILLE, 30C788	61.974	3.0	1.5	1.0	
	GOLD RIVER, 76C285	3.677	3.3	-0.3	2.4	
	KELSEY BAY, 78C145	7.000	3.4	5.5	2.6	
o	CAMPBELL RIVER, 76C267	3.625	2.9	2.3	1.1	`65-`84
o	PORT HARDY	2.369	2.9	3.3	1.1	`64-`84
	CLINTON, 27C580J	883.106	3.6	1.3	1.3	
	PRINCE GEORGE, 17C161H	572.469	4.5	-3.4	1.7	
	ASHCROFT, 16C76J	306.612	4.2	0.2	1.5	
	HOPE, 15C 43J	42.041	4.2	2.5	1.9	
	PRINCETON, 46C570H	652.445	3.9	1.4	1.7	
	CLE ELUM, M 17	581.487	3.1	-0.3	0.9	
o	FULFORD HARBOUR, TGBM 1	1.449	2.9	0.8	0.9	`53-`84
	MALOTT, W20	252.974	3.2	-1.6	1.1	
o	ASTORIA			1.7	1.0	`35-`83
+	ALERT BAY			2.8	1.3	`65-`78

o Permanent tide gauge
+ Subordinate tide gauge

in these regions is determined to be 0.9 mm/yr or less. The lines extending eastward from Puget Sound and northward from Vancouver cross over 1200 m topography and hence may still be affected by residual scale and refraction errors. The dates of the two levelings between Rockport and Malott, 1958.8 and 1982.6, provide only a short time interval within which real tectonic motion could begin to exceed the accumulated uncertainties of the measurements along these lines. In contrast, the time interval between old and new levelings along the line directed north from Vancouver exceeds 50 years, and therefore, the impact of such residual errors should not be great.

However, this latter line is at least partly influenced by magnetic error. Most of the 1983 leveling on this route was done with a Wild NA2 level, but approximately 200 km was levelled in at least one direction with a Zeiss Ni-1. It was not possible to derive a reliable empirical correction for these data, but it is estimated that about 25% of the observed elevation changes may be attributable to magnetic error. However, it is clear that magnetic error is not likely to dominate the 1983 data since the rate and direction of tilting derived from the 1928 and 1958 spirit levelings are consistent with those derived from the 1928 and 1983 levelings. The effect of magnetic errors is also a major concern for leveling lines across central Vancouver Island between Campbell River and Gold River (Dragert, 1987). Although empirical corrections have been applied to Zeiss Ni-1 data from 1976 and 1978, even a small residual error can bias the derived velocity patterns because of the short interval (1976 to 1985) between levelings.

In general terms, the gross velocity pattern shown in Figure 6 exhibits a region of general uplift in the northwest map area and subsidence in the southeast map area. Complicating this simple large-scale pattern is an uplift ridge extending from Neah Bay north to Campbell River (with an accompanying depression in the Straight of Georgia to the immediate east), and a centre of uplift to the west of Princeton in southern British Columbia. The large scale pattern is roughly consistent with vertical deformation generated by the plate tectonic model proposed for this region of the Pacific Northwest. In the northeast map area of Figure 6, the younger (6 Ma), hotter, more buoyant part of the Juan de Fuca plate underlies the coastal margin of southwest British Columbia, and the northwest strike of the coastline establishes orthogonal convergence. Furthermore, the Explorer plate, which underlies much of this region and comprises the youngest oceanic plate, appears to have ceased descending into the upper mantle because of its buoyancy (Riddihough, 1984) and is simply "underplating" the continental margin. The overriding of this hot, buoyant oceanic plate should lead to uplift of the overlying crust. The extension of this uplift region to the south of the inland extrapolated Explorer/Juan de Fuca plate boundary (Pemberton region), although not well-defined by this study, may have a separate thermal origin. Extremely high heat flows have been measured in this young volcanic region 100 km north of Vancouver (Lewis et al., 1988). In the southeast map area, the oldest (9 Ma) and least-buoyant part of the Juan de Fuca plate is converging obliquely with the North American plate and actively descending into the upper mantle beneath coastal Oregon and southern Washington. This is likely to result in less uplift of these continental margin areas compared to the north. However, it is difficult to see how actual subsidence can be generated by this model. Assuming this subsidence is not an artifact of an oceanographic signal, it may imply a north-south flexure of the crust or the existence of extensional stress due to a mild form of back-arc spreading (cf. Rogers, 1985).

The contours defining the region of uplift to the east of Princeton have a much greater uncertainty than the contours in the coastal regions. If this uplift is real, it is too distant to be directly generated by the plate dynamics of the Cascadia Subduction Zone, and some other physical process, possibly thermal, needs to be invoked for its cause. Mathews et al. (1970) used a variety of geological techniques to determine the vertical response of the land to deglaciation in southwestern B.C. and Washington State. It was concluded that vertical motion in the last 5500 years should not be attributed to deglaciation. The ridge of uplift extending north from Neah Bay towards Campbell River is consistent with the transient strain pattern associated with a locked subduction zone (cf. Thatcher and Rundle, 1979). In their models, maximum uplift rates overlie the front edge of the locked portion of the thrust zone, which is at a depth of about 25 km in the Neah Bay region (Clowes et al., 1987). Implications are that the subduction zone is locked to this depth. The location of maximum uplift on the surface will therefore depend on the angle of subduction and not necessarily be a fixed distance from the offshore deformation front. The angle of subduction does change significantly along the Cascadia Subduction Zone (Crossen and Owens, 1987), apparently becoming steeper to the south of the Olympic Peninsula. Although not yet mapped through seismic reflection studies, the dip of the Explorer plate can be expected to be shallower beneath northern Vancouver Island because of its youth and buoyancy. This would be adequate to explain the displacement of the axis of maximum uplift from the outer coastline of Washington to the inner coast of Vancouver Island north of Campbell River.

Acknowledgments. Doug Martin and Bill Stoney of the National Ocean Service helped retrieve, organize, and analyze the data from subordinate tide gauge stations, and provided excellent consultation on the computation of velocities at tide gauges. The Canadian tide data were supplied by Fred Stephenson and W. Rapatz of the Institute of Ocean Sciences, Department of Fisheries and Oceans in Sydney, B.C. Canada.

References

Ando, M., and E. I. Balazs, Geodetic evidence for aseismic subduction of the Juan de Fuca plate, J. Geophys. Res., 84, 3023-3028, 1979.

Balazs, E., and S. Holdahl, Vertical crustal movements in the Seattle-Neah Bay area as indicated by the 1973-4 precise relevelings and mareograph observations. EOS Trans. AGU, 55, 1104-1199, 1974.

Chelton, D. B. and D. B. Enfield, Ocean Signals in Tide Gauge Records. J. Geophys. Res., 91, B9, 9081-9098, 1986.

Clowes, R.M., M.T. Brandon, A.G. Green, C.J. Yorath, A. Sutherland-Brown, E.R. Kanasewich, and C. Spencer, LITHOPROBE - southern Vancouver Island: Cenozoic subduction complex imaged by deep seismic reflections. Can. J. Earth Sci., 24, 31-51, 1987.

Crossen, R.S., and T.J. Owens, Slab geometry of the Cascadia subduction zone beneath beneath Washington from earthquake hypocentres and teleseismic converted waves. Geophys. Res. Letters, 14, 824-827, 1987.

De Jong, S.H., and H.W.F. Siebenhuener, Seasonal and secular variations of sea level on the Pacific coast of Canada. Can. Surveyor, 26, 4-19, 1972.

Dragert, H., A. Lambert, and J. Liard, Repeated precise gravity measurements on Vancouver Island, British Columbia. J. Geophys. Res., 86: pp 6097-6106, 1981.

Dragert, H., A summary of recent geodetic measurement of surface deformation on central Vancouver Island, British Columbia. The Royal Society of New Zealand, Bulletin 24, Proceedings of the International Symposium on Recent Crustal Movements of the Pacific Region, Wellington, N.Z., February, 1984, 29-37, 1986.

Dragert, H., The fall (and rise) of central Vancouver Island: 1930-1985, Can. J. Earth Sci., 24, 689-697, 1987.

Duzois, L. O. R., Precise levelling in British Columbia and Yukon Territory. Geodetic Survey of Canada Publication No. 24, Surveys and Mapping Branch, Canada Department of Mines and Technical Surveys, Ottawa, 238 pp, 1951.

Faucher, F., and S. Blackie, A study on apparent vertical crustal motion on Vancouver Island. Report of the Geodetic Survey of Canada, Ottawa, 1986.

Heroux, P., W. Gale, and F. Faucher, Field test report on the systematic effect of refraction in precise levelling. Proceedings of NAVD Symposium '85, available from National Ocean Service, Rockville, Maryland 20852.

Holdahl, S. R., and R. L. Hardy, Solvability and multiquadratic analysis as applied to investigations of vertical crustal movements. Tectonophysics, 52, 139-155, 1979.

Holdahl, S. R., A model of temperature statification for correction of leveling refraction. NOAA Technical Memorandum NOS NGS 31 (1981).

Holdahl, S. R.: The correction for leveling refraction and its impact on definition of the North American Vertical Datum. Surveying and Mapping, Vol. 43, No. 2, 123-140, 1983.

Holdahl, S. R., Readjustment of leveling networks to account for vertical coseismic motions. Proceedings of International Symposium on Recent Crustal Motions. Tectonophysics 130, 195-212, 1986.

Holdahl, S. R., D. M. Martin, and W. M. Stoney, Methods for Combination of water level and leveling measurements to determine vertical crustal motions. Proceedings of the Symposium on Height Determination and Recent Crustal Movement in Western Europe, Hannover, Fed. Rep. of Germany, 1986b.

Holdahl, S. R., W. E. Strange, and R. J. Harris, Empirical calibration of Zeiss Ni-1 level instruments to account for magnetic errors. Manuscripta Geodaetica 12:28-39, 1987.

Lewis, T.J., W.H. Bentkowski, E.E. Davis, R.D. Hyndman, J.G. Souther, and J.A. Wright, Subduction of the Juan de Fuca plate: Thermal consequences. J. Geophys. Res., (submitted), 1988.

Mathews, W. H., J. G. Fyles, and H. W. Nasmith, Postglacial crustal movements in southwestern British Columbia and adjacent Washington State. Can. J. Earth Sci., 7, 690-702, 1970.

Merry, C. L., and P. Vanicek, Investigation of local variations of sea-surface topography, Marine Geodesy, vol. 7, No. 1-4, 101-126, 1982.

Milne, W. G., G. L. Rogers, R. P. Riddihough, G. A. McMechan, and R. D. Hyndman, Seismicity of western Canada. Can. J. Earth Sci., 15, 1170-1193, 1978.

Nakiboglu, S. M., Global sea level change and vertical crustal motion in Europe. Proceedings of the Symposium on Height Determination and Recent Vertical Crustal Movements in Western Europe, Hannover, Fed. Rep. of Germany, Sept. 15-19, 1986.

Reilinger, R., and J. Adams, Geodetic evidence for active landward tilting of the Oregon and Washington coastal ranges. Geophys. Res. Letters, 9, 401-403, 1982.

Riddihough, R. P., Contemporary movements and tectonics on Canada's west coast - a discussion. Tectonophysics, 86, 319-341, 1982.

Riddihough, R. P., Recent Movements of the Juan De Fuca Plate system. J. Geophys. Res., 89, 6980-6994, 1984.

Rogers, G. C., and H. S. Hasegawa, A second look at the British Columbia earthquake of June 23, 1946. Bull. Seismol. Soc. Am. 68, 653-675, 1978.

Rogers, G. C., Variation in Cascade volcanism with margin orientation. Geology, 13, 495-498, 1985.

Rogers, G.C., Megathrust potential of the Cascadia subduction zone. Can. J. Earth Sci., in press, 1987.

Rumpf, W. E., and H. Meurisch, Systematische Anderungen der Ziellinie eines prazisions Kompensator-Nivelliers--insbesondere des Zeiss Ni-1--durch magnetische Gleich- und Wechselfelder. Federation Internationale de Geometres XVI, Internat. Kongress, Montreux, Schweiz, 1981.

Slawson, W. F., and J. C. Savage, Geodetic deformation associated with the 1946 Vancouver Island, Canada, earthquake. Bull. Seismol. Soc. Am., 69, 1487-1496, 1979.

Thatcher, W., and J.B. Rundle, A model for The earthquake cycle in underthrust zones. J. Geophys. Res., 84, 5540-5556, 1979.

Thatcher, W., and J.B. Rundle, A viscoelastic coupling model for the cyclic deformation due to periodically repeated earthquakes at subduction zones. J. Geophys. Res., 89, 7631-7640, 1984.

Vanicek, P., To the problem of noise reduction in sea level records used in vertical crustal movement detection. Physics of the Earth and Planetary Interiors 17, 265-180, 1978.

Vanicek, P., and D. Nagy, On the compilation of the map of vertical crustal movements in Canada. Tectonophysics, 71, 75-86, 1981.

Wigen, S. O., and F. E. Stephenson, Mean sea level on the Canadian west coast. In: Proceedings of the NAD Symposium 1980, 105-124, Available from the Canadian Institute of Surveying, Box 5378, Station F, Ottawa, Canada K2C 3J1, 1980.

KINEMATICS AND MECHANICS OF TECTONIC BLOCK ROTATIONS

Amos Nur[1], Hagai Ron[2], and Oona Scotti[3]

Abstract. Large portions of the earth's crust are broken by dense *sets* of parallel or subparallel faults which are organized in *domains*. Analysis of structural and paleomagnetic data suggest that when such domains are subject to tectonic shearing they deform by distributed fault slip and block rotations, rather than by uniform straining. Many such domains have been recognized in the Western U.S., and in California and Nevada in particular. Mechanical considerations of stress, strength and friction reveal under what conditions new fault sets must form when these rotations are sufficiently large (25°-45°) leading to domains of *multiple sets*. Several domains of multiple fault sets have by now been recognized in localities in California and Nevada.

In the past many studies of rotations were limited to separate structural, or paleomagnetic data. However by combining paleomagnetic, structural geology, and rock mechanics data, we are able to explore the validity of the block rotation concept and its significance in much greater detail than ever before. Our analysis here is based on data from (1) Northern Israel, where fault slip and spacing is used to predict block rotation; (2) the Mojave Desert, with well-documented strike-slip fault sets, organized in at least three major domains. A new set of faults trending N-S may be in the process of formation here; and (3) the Lake Mead, Nevada, fault system with well-defined sets of strike-slip faults, which, in contrast with the Mojave region, are surrounded with domains of normal faults; and (4) the San Gabriel Mountains domain with a multiple set of strike-slip faults.

Block rotations can have profound influence on the interpretation of geodetic measurements and the inversion of geodetic data, especially the type collected in GPS surveys. Furthermore, block rotations and domain boundaries may be involved in creating the heterogeneities along active fault systems which are responsible for the initiation and termination of earthquake rupture.

---

[1]Department of Geophysics, Stanford University, Stanford, California 94305

[2]Institute for Petroleum Research and Geophysics, P.O. Box 2286, Holon, Israel

[3]Department of Geophysics, Stanford University, Stanford, California, 94305

Copyright 1989 by
International Union of Geodesy and Geophysics
and American Geophysical Union.

The Problem: Deformation of the Crust with
Distributed Fault Sets

More than any other active plate boundary region, Western North America has been investigated in great detail for decades. Due to the relatively dry climate, exceptional exposures of rocks and geologic structures exist in this region, which have attracted many geological and geophysical studies. The potential damage of future large earthquakes has attracted also a great deal of geophysical research on the distribution and activity of faults throughout this region. It is not surprising therefore that several major and classical geological and geophysical concepts of crustal deformation have developed here, including the idea of very large offsets on strike-slip faults, the earthquake rebound theory, crustal extension in the Basin and Range, and the concepts of pull-apart basins accreted terranes, and large crustal rotations.

Despite this past and ongoing research activity much confusion still exists as to the details of crustal deformation in the Western United States. This is due, more than anything, to the fact that much of the deformation here is distributed over a huge network of numerous large and small faults oriented in many different directions, and active more or less simultaneously. In this paper we review a new integrated approach to the kinematics and mechanics of the deformation of crustal regions which contain distributed fault systems, with particular emphasis on selected portions of Western USA.

There are four specific features of distributed fault systems which are directly relevant to the problem we wish to consider here: (1) the existence of *sets* of many faults, particularly sets of parallel strike-slip faults; (2) the organization of these sets in *domains*; (3) the systematic tectonic *rotations* of blocks between faults within sets in a given domain; and (4) the presence within some domains of several intersecting fault sets, or *multiple sets*.

(a) Fault sets. Throughout the Western United States, and in many other tectonic regions in the world, sets of roughly parallel strike-slip faults are found (fig. 1). These faults within a set, are active roughly simultaneously. Although such sets are to be expected as part of major fault systems, e.g. the San Andreas system, sets are also common away from the San Andreas, distributed as they are throughout the Mojave Desert, and in large portions of the Basin and Range Province. Some questions which arise are: why do these sets exist, and what controls the spacing between faults within these sets?

(b) Domains of fault sets. Strike-slip faults tend to occur in domains of left- or right- slipping faults in the same tectonic environment (Freund, 1970; 1974), e.g. the two domains of the Mojave Desert (fig. 2) and northern Israel. Unlike classic

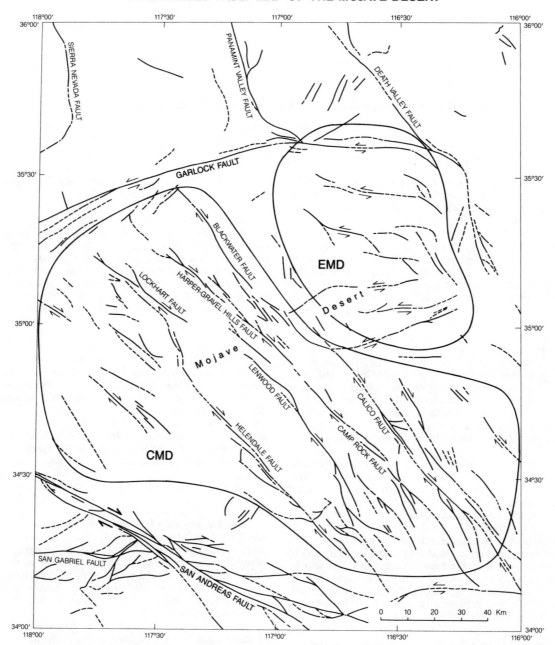

Fig. 1. Details of the central Mojave Desert (CMD) and eastern Mojave Desert (EMD). Each domain consists of roughly parallel fault sets of consistent sense of slip, which is expected to be directly associated with block and fault rotations: counterclockwise rotation with right-lateral slip in the CMD, and clockwise rotation with left-lateral slip in the EMD.

models (Billings, 1972; Anderson, 1951) which predict slip on conjugate faults, deformation actually seems to occur in "conjugate" domains where one domain may show left-lateral slip, and a neighboring domain may show right-lateral slip. This pattern has also been recognized in many distributed strike-slip systems worldwide. Why these domains form, what controls their dimensions and how they are related to slip and overall deformation are key questions still to be answered.

(c) Block and fault rotation. One of the most important aspects of fault sets in domains is the rotation of the blocks between the faults when they slip, and the consequent rotation of the faults themselves (Freund, 1970a, 1974; Garfunkel,

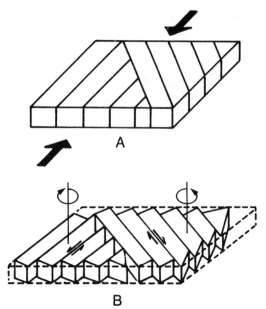

Fig. 2. A 2-D model which illustrates the simultaneous activity of strike-slip displacement and rotation of the faulted blocks. (a.) The initial configuration; (b.) after deformation. The set of left-lateral faults rotate clockwise and the set of right-lateral faults rotates counterclockwise.

1974). That these rotations take place has by now been established by comparing paleomagnetic with detailed structural data-namely fault slip and fault spacing (Luyendyk et al., 1980, Ron et al., 1984) with surprisingly good agreement in areas where both data were obtained.

Simple considerations (Freund, 1970a) show that block and fault rotations associated with fault slip imply that sets with left-lateral slip undergo clockwise rotation, whereas right-lateral sets undergo counterclockwise rotation (fig. 2). These considerations suggest that fault sets and their domain geometry are time variable. An important question which arises here is: How do domain boundaries accommodate the differences in rotations between two adjacent domains?

(d) Multiple sets. Finally there is an important mechanical aspect of distributed fault sets and their rotations. As faults in a set rotate away from the direction of optimal slip, they eventually must lock up in directions relative to the stress field in which slip is inhibited. Further deformation must then be accommodated on newly formed faults which are more favorably oriented to the stress. The creation of these new sets leads to domains with multiple sets of faults, with older sets offset by younger ones. Multiple sets have been recognized in several domains in Western North America and elsewhere. The obvious questions which arise: under what conditions do new sets develop, and how are they related geometrically (length, spacing, termination) to the older sets?

The Model: Block and Fault Rotations

We have investigated two aspects of crustal deformation in regions with pervasive sets and multiple sets of strike-slip faults by the combination of (Ron et al. 1984; Nur et al. 1986) four separate fields (Ron et al. 1986) (1) structural and field geology; (2) paleomagnetism; (3) rock mechanics; and (4) seismicity. The synthesis of these inputs was based on a two-pronged analysis of the tectonics of regions with fault sets: (1) the kinematics, which involve the description of block and fault rotations, in sets and domains, and the role of fault spacing, slip and directions of the faults; and (2) the mechanics, which involve fault and block deformation and rotations subject to stresses and constrained by the frictional and fracture strength properties of crustal rocks.

Kinematics

It has long been recognized (Freund, 1970a) that fault blocks in strike-slip tectonic domains must progressively rotate on vertical axes as the overall strike-slip motion continues. Two direct consequences of this deformation mechanism are that (1) slip on each of the faults within a domain must be related to the rotation of the blocks bounded by these faults, and that (2) the faults themselves must also rotate because they are the boundaries of the blocks (fig. 2). These ideas were applied to strike-slip tectonics in Eastern Iran (Freund, 1970a), New Zealand (Freund, 1971), the Dead Sea transform (Garfunkel, 1970), Mojave Desert, California (Garfunkel, 1974), Southern California (Luyendyk et al, 1980), and Southeast Sinai (Frei, 1980). The actual geometric relations involved in this rotation process were analyzed by Freund (1974), Garfunkel (1974), Luyendyk et al. (1980), and MacDonald (1980), who showed how the rotations of blocks and strike-slip displacements are two interrelated and contemporaneous aspects of a single deformation process. Specifically if the blocks are rigid, then the model predicts a simple quantitative relationship between fault spacing, slip, and amount and sense of block rotation. Thus in areas where fault spacing and net slip can be determined geologically, paleomagnetic measurements may be used to test the sense and amount of rotation predicted by the model. One of the most basic aspects of the rigid fault and block rotation model (Freund, 1970a, b, 1974; Garfunkel, 1974) is that the sense of block rotation must be opposite to the sense of the fault slip, with left-handed slip associated with clockwise rotation, and right-handed slip with counterclockwise rotation (fig. 2).

The geometrical relation (fig. 3) between the displacement d along a fault (positive when right-lateral), the width w of the

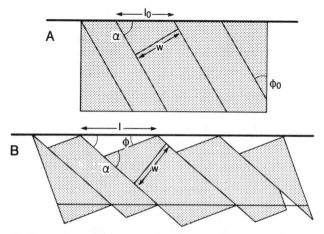

Fig. 3. Fault set kinematics. Geometrical relations among spacing between faults w, their initial orientation α, initial length l_0 to slip d, and rotation ϕ.

TABLE 1. Reported Block Rotations Inferred from Paleomagnetic and Structure Data

Location of rotated blocks	Magnitude and sense of Rotation	References
Iran-Sistan	± 45°(S)	Freund, 1970
Jordan Valley-Israel	-25, -72° (P)	Nur and Helsley, 1971
Mojave-California	-30° (S)	Garfunkel, 1974
Southern California	+45° to +90° (P,S)	Luyendyk et al., 1980
Mojave-California	-16° (P)	Morton and Hillhouse, 1983
Southwest Arizona	-14° (P)	Calderone and Butler, 1984
Imperial Valley, California	+30 to +70° (Sd)	Terres and Sylvester, 1981
Morro Rocks-Islay Hill complex, California	+49° (P)	Greenhouse and Cox, 1979
Eastern Klamath belt, northern California	+83 to +116° (P)	Fagin and Gose, 1983
Southwest Washington	+20 to +65° (S,P)	Wells and Coe, 1985
Northern Israel	+23 to +36° (S,P) -24 to -53°	Ron et al., 1984
Hermon and Lebanon, northern Israel and Lebanon	-59° (P,S)	Ron and Nur, in press
Rio Grande rift, New Mexico	+12, -16° (P)	Brown and Golombek, 1985
Dixie Valley, west-central Nevada	-36° (P)	Hudson and Geissman, 1985
Lake Mead, Nevada	-27° (P)	Ron et al., 1986
San Gabriel region, California	+53° (S,P)	Terres and Luyendyk, 1985
Northern Channel Islands, California	+74° (P)	Kamerling and Luyendyk, 1985
Western transverse ranges	+36 to +92° (S,P)	Hornafius, 1985

Note: P = paleomagnetic; S = structure; Sd = surface deformation;
- = counter-clockwise; + = clockwise

faulted block, the inital angle α between the faults and the boundary of the domain, and block rotation φ (positive when counterclockwise) is given by

$$d/w = \frac{\sin\phi}{\sin\alpha \cdot \sin(\alpha-\phi)} \qquad (1)$$

The relative elongation l/l_0 of the faulted domain parallel to its boundary, is given by

$$\lambda = l/l_0 = \frac{\sin\alpha}{\sin(\alpha-\phi)} \qquad (2)$$

The deformation within a single fault domain is simple shear.

The block rotation model can be rigorously tested in areas where domains of strike-slip faults exist and where the blocks are fairly rigid and where paleomagnetic declinations can be measured to reveal possible horizontal rotation about vertical axes. If the model is correct, we expect to find agreement using equation (1) between structural rotations computed from the amount and sense of fault slip and and fault spacing as measured in the field, and rotation amount and sense obtained from paleomagnetic declination measurements (Ron et al. 1984).

Mechanics

In many published cases, the rotations inferred from paleomagnetic data are large, exceeding 45°; some values approach 100° of block rotation (table 1). If the faults rotate with the blocks, as envisioned by the block rotation model, then a major question arises: How much can a fault plane rotate away from the optimal direction of shearing before it cannot slip any more? If the magnitude of such rotation is restricted, then the kinematic model as described above must be modified accordingly.

An obvious constraint on the amount of fault rotation is imposed by the mechanical condition of faulting, namely that shear stress on the plane of the fault must exceed the shear resistance to slip along the fault. From extensive laboratory studies, it has been established that the shear stress τ required for sliding on a preexisting fracture is controlled by the fracture cohesive strength S_1, and the effective normal stress τ_0 acting on the fracture

$$\tau = S_1 + \mu\sigma_0 = S_1 + \mu(\sigma-p) \qquad (3)$$

where μ is the coefficient of friction, σ is the normal stress, and p is pore pressure. If we consider a fault originally formed at

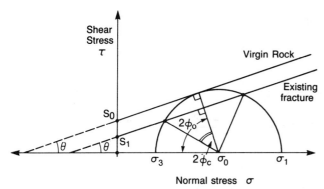

Fig. 4. Mohr circle representation of the state of stress in rock with existing fracture: $\theta = \tan^{-1}\mu$ is angle of friction, S_0 and S_1 are cohesive strength of virgin rock and fracture, respectively, and ϕ_c is largest permissible angle between existing fracture and direction in which new shear fracture will develop if further rotation takes place.

the optimal direction of failure ϕ_0 relative to the maximum stress as shown in the Mohr diagram of figure 4, it follows that as deformation and rotation proceed, the shear stress acting on the fault plane decreases, and the normal stress increases. Beyond some critical angle of rotation ϕ_c it becomes easier for a new fault to form in the rock mass with cohesive strength S_0 in direction ϕ_0, relative to the stationary direction of maximum stress, rather than continue sliding on the preexisting one, now oriented in the direction ϕ_c from ϕ_0.

The critical angle ϕ_c is given by (Nur et al., 1986)

$$\phi_c = \frac{1}{2} \cos^{-1}\left[1 - \frac{(1 - S_1/S_0)}{1 + (\mu\sigma_0/S_0)}\right] \quad (4)$$

where S_0 and S_1 are the cohesive strengths of the virgin rock mass and the preexisting fracture respectively, $\theta = \tan^{-1}\mu$, and σ_0 is effective overburden pressure.

Equation (4) shows that there is a mechanical limit on fault rotation under a stationary stress field. Apparently, the magnitude of the permissible rotation ϕ_c is dependent on the difference between the cohesive strength S_0 of the virgin rock and that of the preexisting faults S_1, the coefficient of friction $\mu = \tan\theta$, and effective overburden pressure σ_0 (or equivalently, depth).

The results of equation (4), plotted in figure 5, show that under conditions of constant stress field the kinematic model must be modifed, as shown in figure 6, to include the appearance of a new set of shear faults when rotation is sufficiently large. The values of obtained from equation (2) show that the angle to which a fault set can rotate before a new set must appear to accommodate further block rotation is in the range of 25° to 45°. For the extreme cases $\mu = 0$ or $\sigma_0 = 0$, we obtain the maximum possible angle of rotation of $\phi_c = 45°$, which is the absolute upper limit of block rotation which can be accommodated by one set of faults. In situ, the angle ϕ_c should be the angle between the direction of the faults in the older locked set, and the orientation of the faults in the newer set, which offsets the older one.

Field Evidence

Northern Israel. Probably the first suggestion that large spatial variation of paleomagnetic declinations may be due to tectonic rotations about vertical axes was made by Nur and Helsley (1971) as a result of a field study in Northern Israel in 1967. These results revealed large declination anomalies which have suggested the possibility of a rigorous test of the block tectonics model. This test was carried out by Ron et al. (1984) in Northern Israel (fig. 7), where good structural data on fault spacing and slip were available and good paleomagnetic data yielded reliable and sufficiently accurate declination values.

The results of this study-both structural and paleomagnetic-revealed several domains of fault sets, some with clockwise and left-lateral fault slip, others with counterclockwise and right-lateral slip, as predicted by the model of Freund (1970). Figure 8 shows a summary of the structurally observed rotations vs. the paleomagnetically measured ones, with values of the rotations in the various domains ranging from 0° to 54°. The results clearly show that the magnitude and sense of block rotations as predicted from the structural data agree well with values obtained from the independent paleomagnetic determinations, suggesting that the rigid fault rotation process may thus be typical of such systems in general.

Multiple sets. There are a few in situ cases in which the angle ϕ_c between the directions of faults in an older and a newer fault set within one domain have been determined: The values of these angles range between 25° to 45°, well within the predicted range of mechanically permissible fault rotations. These values are, however, significantly lower than many of the total block rotations determined paleomagnetically (table 1). According to our model block rotations larger than 45° must involve two or more sets of rotating faults; and rotations over 90° require three sets or more. One of the examples of

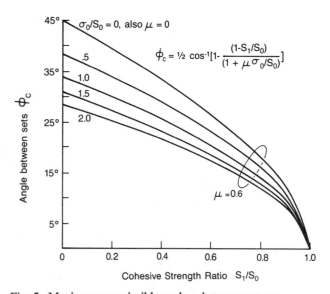

Fig. 5. Maximum permissible angle ϕ_c between sets or rotating shear faults as function of cohesive strength of rock S_0 and fault S_1, and effective overburden pressure σ_0 for $\mu = 0.6$ and $\mu = 0$.

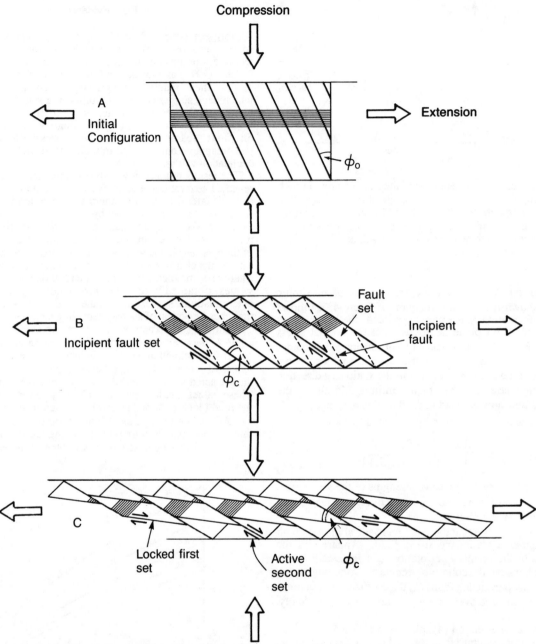

Fig. 6. Modified block model with critical angle ϕ_c given in figure 5 showing new fault set required to accommodate block rotation greater than 45°.

table 1 was recognized by Freund (1970) in the Sistan, Iran, region, where two sets of strike-slip faults 40° to 45° apart were found to have accommodated the rotation of crustal blocks. Three sets may be involved in north-central Iceland (Young et al., 1985), and in the Snake Range, Nevada (Gans and Miller, 1983). Although the rotation in the Snake Range is associated with normal, not strike-slip, faulting, the model is similarly applicable. It is noteworthy that the angles between the sets as reported by Gans and Miller are again within the range of values permissible by the model. Proffett (1977) has similarly identified at least two multiple normal fault sets in the Yerington, Nevada, area with an angle between sets of around 40°.

The applicability and usefulness of the block tectonics approach in the Western U.S. can be evaluated by considering two areas--the Mojave Desert, and the Lake Mead, Nevada, fault zone--where comparisons between structurally inferred rotations, derived from fault geometry and slip, and

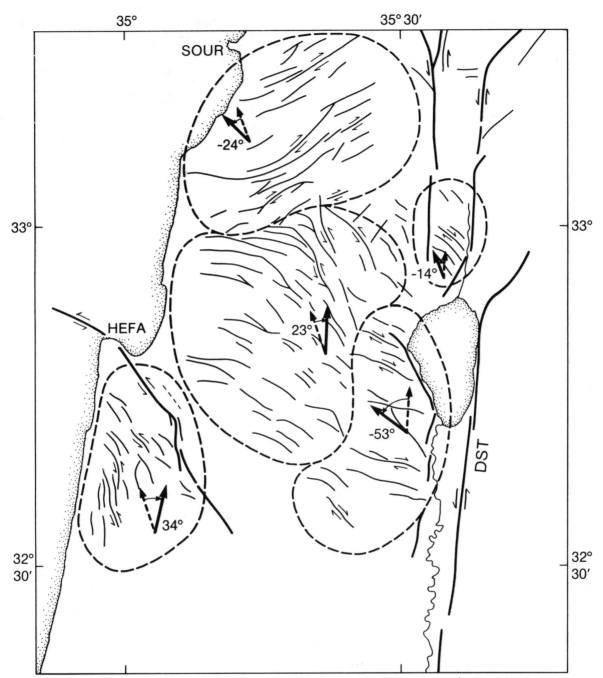

Fig. 7. Fault domains in northern Israel, showing fault sets, directions of slip, and paleomagnetically determined rotations in these domains. DST-Dead Sea transform.

paleomagnetically derived rotations, computed from available declination measurements can be made.

<u>The Central Mojave domain</u>. The late Neogene crustal deformation in the Mojave Desert is predominantly accommodated by sets of NW-trending right-lateral strike-slip faults and E-W-trending left-lateral strike-slip faults (fig. 1) (Dibblee, 1961; Garfunkel, 1974; Dokka, 1983). On the basis of structural data, Garfunkel (1974) suggested that deformation in the Mojave is accommodated by contemporaneous right-lateral strike-slip faulting and counterclockwise (CCW) block rotation.

The anticipated rotation in the Central Mojave domain is counterclockwise. The magnitude of block rotation is estimated to be around 15° (CCW) using Dokka's (1983) fault

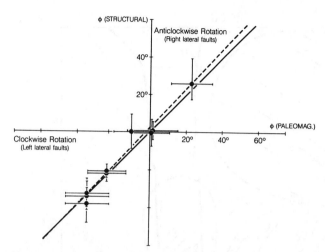

Fig. 8. Comparison between structurally derived and paleomagnetically determined rotations in northern Israel (see fig. 7). Solid line represents perfect agreement; dashed line is linear best fit to data.

slip values. Paleomagnetic data from this region (unpublished, Morton and Hillhouse, 1983; McFadden et al. 1987) show about 15° CCW rotation of the Mojave blocks during the last 6 Ma.

Seismic, active surface deformation, and geodetic data, may also be consistent with crustal deformation being accommodated by the block rotation model even at present. In 1975, the Galway Lake earthquake (M = 5.2) (fig. 9) was accompanied by a 6.8 km long surface rupture with right-lateral displacement (Hill and Beeby, 1977) on a previously unmapped fault trending N10°W. Remarkably this displacement did not occur along the established nearby Emerson Fault, which trends N 45° W. Sauber et al. (1986) concluded that the preferred nodal planes for five out of nine recent earthquakes in this area are also oriented 20° to 35° east from the general strike of the Mojave faults, roughly in the direction of the Galway rupture direction, although they consider this direction of faulting to be "secondary." Furthermore, the fault plane solutions for the larger shocks of the 1979 Homestead Valley earthquake sequence (Hutton et al., 1980) show nodal planes trending in N 4° W, again 30° to 40° away from the strike of the more developed, presumably older Mojave fault trend (Stein and Lisowski, 1983). It is remarkable that both ruptures together delineate a single line which combined suggest a single fault at least 25 km long.

We suggest that the direction of faulting in the central Mojave domain of the Galway Lake and Homestead Valley earthquakes might very well be part of a developing new fault set, which is gradually replacing the older, now rotated out of favor strike-slip faults in the central Mojave domain. This interpretation is consistent with the direction of the compressive stress in this area which is horizontal, at approximately N 30° E to N 45° E as given by Zoback and Zoback (1980). Accordingly the older Mojave faults are thus unfavorably oriented relative to this stress direction, whereas the Galway Lake-Homestead Valley faulting direction is mechanically quite favorable.

The Lake Mead strike-slip fault system. The Lake Mead fault system (LMFS) consists of a few major NE-trending left-lateral strike-slip faults. The accumulated offset across these faults is about 65 km, as obtained from offset of Late Neogene volcanic rocks (Anderson, 1973; Bohannon, 1979). Based on the ages of the displaced rocks of the Hamblin-Cleopatra volcano (Anderson et al., 1972) strike-slip faulting here post dates 11 Ma.

A detailed study of the geometry and nature of faulting in the LMFS was carried out by Angelier et al. (1985), in the Hoover Dam area. They conclude that two sets of strike-slip faults are present: A set of major NE trending left-lateral faults, which are typically 50 to 150 km long, and a second set with faults trending roughly NW. The second set is divided into two fault groups: right-lateral strike-slip and normal dip-slip faults. The NW-trending right-lateral faults are typically 5 to 25 km long, and roughly conjugate to the NE set.

Paleomagnetic data from the Hamblin-Cleopatra volcano were obtained (Ron et al., 1986) from 22 cooling units of volcanic rocks with K-Ar ages by Anderson et al. (1972) of 11.3 Ma and 12.7 Ma. The paleomagnetic results reveal little if any rotations about the horizontal axis, corresponding to the lack of tilting associated with normal fault slip. In contrast, the paleomagnetic declination anomalies yielded large horizontal counterclockwise rotation of -29.4 ± 8.5 about the vertical axis. This counterclockwise sense of rotation is inconsistent with the block rotation model with the left-lateral slip on the NE-trending major fault set. In contrast, this sense of rotation is to be expected if the associated slip took place on the shorter pervasive NW-trending right-lateral faults of the LMSZ. Our interpretation of the results is shown in figure 11, in which the major NE fault coincides with the fixed boundary direction of this domain-which does not rotate-with rotation totally associated with the conjugate NW faults.

Supporting evidence for this interpretation comes from current active faulting (assuming active faulting is tectonically comparable with the late Miocene in the LMFS). According to Rogers and Lee (1976) the majority of earthquake epicenters in the LMSZ are associated neither with the smaller NW trending faults nor with the larger NE ones, but rather with younger and shorter, N-S-trending fault segments (fig. 10). Focal mechanisms for these earthquakes reveal right-lateral strike slip motion on these faults (Rogers and Lee, 1976).

Figure 10 presents a model for the fault geometry, the sense of horizontal slip, and the nature of block rotation in the LMSZ, beginning about 11 m.y. ago and still in progress now. Left-lateral shear in Miocene time caused local right-lateral horizontal displacements on a local set of faults initially trending N-S. These right-lateral displacements lead to an approximately 30° counterclockwise rotation of the blocks and their bounding faults, resulting in their present NW direction in which they became locked (Nur et al. 1986). Assuming that the stress orientation remained constant during the past 11 m.y. (Zoback and Zoback, 1980; Carr, 1984), subsequent deformation by right-lateral shear has shifted to a new set of more favorably oriented N-S faults which are active today.

This interpretation provides a link between the structural, paleomagnetic and seismic data and integrates them into a single continuous process of older and current crustal deformation by strike-slip and block rotation. Although the results are specific to the LMSZ, the model by analogy should be equally applicable to other Nevada strike-slip fault systems. Because this generalization could have profound impact on our

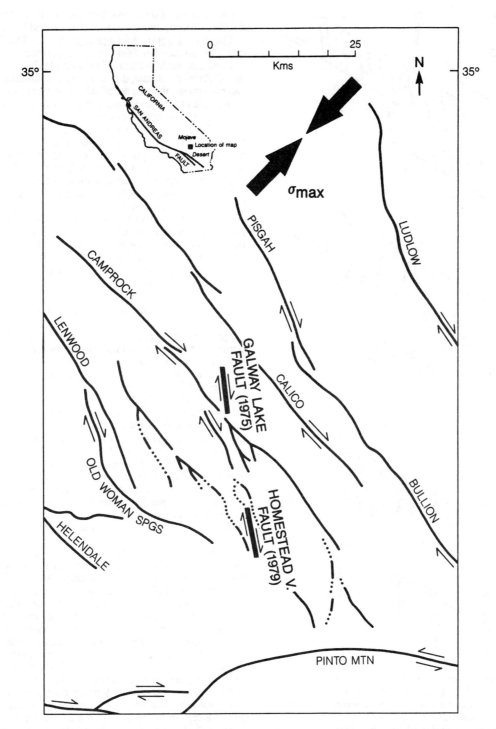

Fig. 9. The formation of a new fault set. The faulting directions associated with the 1975 Galway Lake earthquake and the 1979 Homestead Valley earthquake sequence. Note their azimuths which are distinctly different from the older, well-established central Mojave faults, and their alignment along a single trend. Shown also is the estimated direction of the regional σ_{max}.

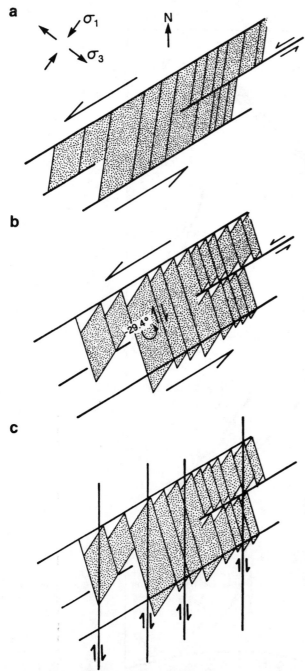

Fig. 10. Simplified geometric model for deformation of the left-lateral Lake Mead fault system by right-lateral strike-slip faults and counterclockwise rotation. Direction of elongation is northwest so that the major NE-SW fault have not rotated. (a.) Stage I. Slip on the major NE-trending faults; (b.) Stage II. Slip and rotation on the shorter NW-trending fault; the original orientation of these faults, based on the measured rotation was N-S; (c) Stage III. Slip and seismicity on the active N-S-oriented faults. The slip was presumably transferred to these faults after the NW-trending set became locked due to their rotation.

understanding of the active tectonics of important portions of the Great Basin, much more work is needed in the future.

The San Gabriel Mountain domain. The formation of multiple strike-slip fault sets, associated with clockwise rotation can explain some of the complexity of faulting in the San Gabriel Mountain domain (fig. 11) Dibblee (1961) long ago recognized that the set of subparallel left-handed strike-slip faults here imply clockwise rotation of both blocks and faults. Subsequent paleomagnetic results by Terres and Luyendyk (1985) revealed such rotation, of about 50° to 55°. As implied by the mechanical analysis (Nur et al., 1986) such rotations may be larger than could be accommodated by a single fault set. As shown in figure 11 (after Carter, 1982), there is good evidence for the presence here of an older strike-slip fault set, which is cut and offset by the younger set. According to the block rotation model, this older set, oriented now roughly NE, has rotated gradually clockwise, until it became locked due to the increasing normal stress and decreasing shear stress, with further deformation being accommodated by the new set of faults. The observed angles between the two sets is in the range of 30° to 40°, in good agreement with the predicted value of Nur et al. (1986).

Relevance to Geodynamics

Scale dependence and mechanics. The block rotation concept may be important for the interpretation of crustal deformation data such as obtained from repeated geodetic measurements. The process of block rotation implies, for example, that strain, fault slip, and crustal deformation must generally be scale-dependent. On the scale of a few kilometers, deformation is controlled by fault spacing; on the scale of a few tens of kilometers, deformation is controlled by fault domains; and on a scale of a few hundreds of kilometers, deformation is controlled by regional tectonics and plate motions. For example, a local domain can have left-lateral slip and clockwise rotation in an overall region with right-lateral and hence counterclockwise rotations. Furthermore, faults must generally not be in directions of optimal failure and are therefore ambiguous regional stress indicators.

In order to incorporate the roles of fault sets and domains in crustal deformation and evaluate their importance, their mechanical role must be understood in some detail. For example rock strength and frictional properties in two and three dimensions must be specifically considered in the formation of fault sets, domains, and domain boundaries. These properties may also control the formation of multiple fault sets of a given type (strike-slip, normal, thrust) or mixed types (strike-slip and thrust in the Transverse Ranges, strike-slip and normal in the Basin and Range). Consequently slip on a single fault, say during a single earthquake, such as the Galway Lake or the Homestead Valley events, does not reliably reveal the regional deformation field. Similarly, the deformation of a given domain does not mimic regional deformation: For example, the overal deformation of the Mojave region is the sum of the deformation of its separate domains. Clearly the strain of the central domain differs markedly from the strain of the eastern domain.

Stress-strain relation for a single domain. As slip on and rotation of the faults of a single domain set progress with time, the local shear stress required to strain the domain must change, until, as shown earlier, it becomes more favorable for a new set of faults to be activated. The Mohr Circle

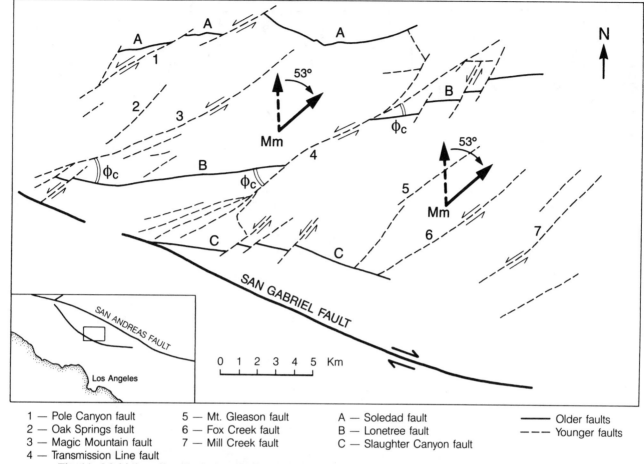

1 — Pole Canyon fault
2 — Oak Springs fault
3 — Magic Mountain fault
4 — Transmission Line fault
5 — Mt. Gleason fault
6 — Fox Creek fault
7 — Mill Creek fault
A — Soledad fault
B — Lonetree fault
C — Slaughter Canyon fault

——— Older faults
- - - - Younger faults

Fig. 11. Multiple strike-slip fault sets in the San Gabriel Mountains (after Carter, 1982). Note the younger strike-slip NE-trending strike-slip faults offsetting the older E-W-trending strike slip. Both sets have the same left-handed sense of motion and are therefore not conjugate sets. The paleomagnetically determined clockwise rotation of 53° (Terres and Luyendyk, 1985) is consistent with the observed left-handed slip.

representation in fig. 12 shows the relation between the normal shear stress τ_f on a representative fault out of a given set. The fault is oriented α degrees from the direction of maximum stress. The angle ϕ gives the rotation of the initial fault from α with $0 \leq \phi \leq \phi_c$, where ϕ_c is the maximum rotation which is mechanically permissible (see fig. 4). The shear stress acting on the fault which has rotated α from ϕ_0 is, as seen in fig. 12, given by

$$\tau_f = \tau_i \frac{1}{\cos 2\phi} \qquad (5)$$

where τ_i is the shear stress for the initial orientation of the fault set, and ϕ ($0 \leq \phi \leq \phi_c$) is the magnitude of block rotation.

The extensional strain ε associated with the rotation is given ϕ by (Freund, 1974)

$$\varepsilon = \frac{l - l_0}{l_0} = \frac{\sin \alpha - \sin (\alpha - \phi)}{\sin (\alpha - \phi)} \qquad (6)$$

where l is the domain length, l_0 is the initial length (when rotation is $\phi = 0$), and α is the initial orientation of the faults relative to the direction of shortening. Using equations 5 and 6 it is now possible to obtain the stress-strain relation for a domain through the common parameter of block rotation ϕ. This relation is obviously non linear, as shown in fig. 13. The relation depends on the magnitude of the optimal shear stress τ_0 (which in turn depends on the coefficient of friction μ and cohesive strength S_1), and on the orientation α of the initial fault set undergoing rotation.

When α is equal or greater than ϕ_0, $\alpha \geq \phi_0$ the domain system exhibits strain hardening in which larger shear stress increments are required per increment of strain as the blocks rotate. However, when $\alpha < \phi_0$, the system exhibits strain softening (fig. 13b) in which less stress is required to cause a slip increment as rotation progresses. This softening behavior of a domain implies a mechanical instability, which might be related to the spatial distribution of seismicity in domains.

From fig. 12 it is easy to recognize that as the fault rotates, the change of shear stress τ_f on the fault itself differs from the

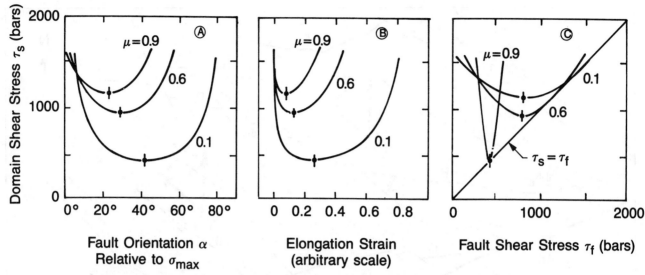

Fig. 12. The changes of stresses with rotation represented by a Mohr circle diagram. Two cases are considered: (a.) when the angle α between the direction of τ_{max} and the fault is larger than the direction of optimal failure ϕ_0, $\alpha > \phi_0$ rotation requires an increase in fault shear stress τ_f relative to domain shear τ_s leading to strain hardening; (b.) when $\alpha < \phi_0$ rotation involved a decrease in τ_f relative to τ_s, leading to strain softening.

domain shear stress τ_s acting on the domain as a whole, and that this difference depends on the rotation ϕ. In fig. 13c, we show the relation between these two shear stresses. The plot further emphasizes the unstable nature of domains in which the direction of the rotating faults α is closer to the direction of shortening than ϕ_0.

Crustal deformation of a cluster of domains. In figure 14 we illustrate one of the problems which can arise in the interpretation of geodetic crustal deformation data when strike slip type block rotation is involved, especially when measurements of azimuth changes are concerned. Consider first two geodetic stations, A and B, within a single domain (fig. 14a). As the domain deforms each station undergoes rotation of the same sense, as would be the case for continuous uniform deformation within the domain. However if station A is in one domain, and station B in a second (fig. 14b), a major difference from the homogeneous case arises, because station B, rotates counterclockwise, whereas station A rotates clockwise. Consequently if the fact that the geodetic survey crosses a domain boundary, is not included in the analysis and the interpretation of the geodetic data, large and fundamental errors can be introduced. In the example here the two rotations would tend to cancel one another, leading to totally erroneous estimates of shear strain.

Relevance to Earthquake Prediction

Although the concept of block rotations is directly applicable to the understanding of distribution faulting, it is not immediately obvious whether and how it might also be related to earthquake prediction. Two specific applications are worth exploring: the initiation and arrest of earthquake rupture, and the prediction of which fault will slip.

As suggested by many investigators, irregularities in the geometry of a fault or fault system may be the cause for either or both the initiation of earthquake ruptures and their arrest. Some models envision that rupture begins at a point of low strength or resistance to slip and is stopped at strong points, or barriers. Other models envision that rupture actually begins at high strength points, and spreads into weak, or compliant regions where it is arrested. Regardless of the model, it is apparent that large heterogeneities of fault strength or normal and shear stresses are required, and that such heterogeneities may be related to geometrical irregularities along faults and fault systems.

As pointed out by e.g. Freund and Merzer (1976) and Nicholson et al. (1986), the rotation of blocks within fault systems, and the rotation of the fault strands themselves may be directly related to the location and creation of heterogeneous normal and shear stresses along active faults. The study of the details of domains of block rotations and their development in time and space may consequently be significant for the understanding of both rupture initiation and stopping, and the prediction of sites for large future earthquakes.

The concept of blocks and their role may also be important in predicting the particular fault on which a future earthquake will occur. This question is of special interest because the most successful earthquake prediction to date-the Haicheng, china, earthquake of February 5, 1975--involved the correct prediction of the time and the approximate location of the event, but failed to predict the fault on which this earthquake occurred. The epicenter instead of alling on the major through-going NW right-handed fault, actually fell on either a conjugate left slipping fault, or a fault in a conjugate set or domain. The stress required to rupture this fault may have been unusually high, due to the unfavorable orientation of the earthquake fault, so that the associated precursors were unusually strong.

A second example is the $M \approx 5.0$ Mount Lewis earthquake of 1985 east of the San Francisco Bay, California. Although

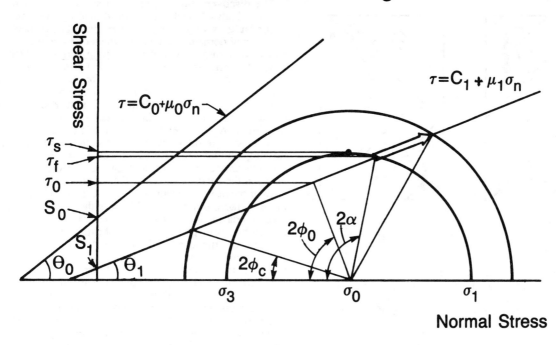

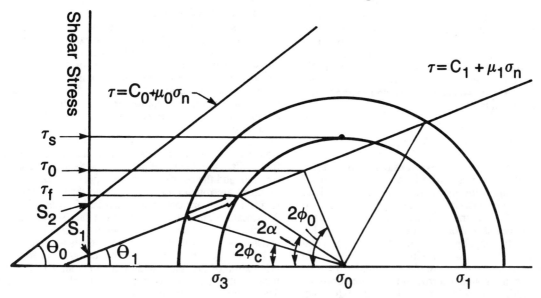

Fig. 13. Relation between domain shear stress τ_s and (a.) fault orientation relative to τ_{max}; (b.) magnitude of strain; and (c.) fault shear stress τ_f, for a range of coefficient of fault friction $\mu = 0.1, 0.6,$ and 0.9. The tick marks on each curve indicate the corresponding values when fault are oriented in the optimal shear direction ϕ_0. Note that when $\alpha > \phi_0$ domain stress increases with increasing fault rotation, elongation and fault shear stress τ_f In contrast, when $\alpha < \phi_0$, domain stress τ_s decreases with increasing rotation, strain, and fault shear stress τ_f.

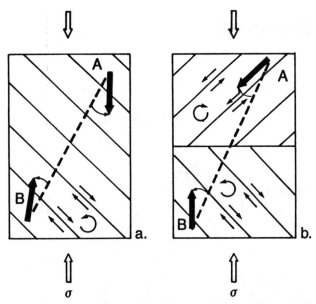

Fig. 14. Sketch illustrating the influence of block rotation and domains on azimuthal changes, as determined from geodetic measurements. Heavy arrows indicate original lines connecting stations A and B. Light arrows indicate sense of fault slip.

the main strands of the San Andreas in this area trend NW-SE, and show right-handed slip, many active north-trending faults are present, possibly due to the presence of a complex of pull-aparts and pushups (Aydin and Page, 1984; Aydin and Nur, 1982). The Lewis earthquake fault plane solution indicates right-handed slip on one of these N-S-trending faults. The block rotation model (fig. 2) provides an explanation for this puzzling earthquake, as an event on one of the faults in the N-S-trending strike-slip set of an east bay fault domain. This domain is bounded by the two major San Andreas strands, the Calaveras and Hayward faults.

Some Outstanding Problems

The concepts and results which we considered so far form the basis we believe for a unified approach to crustal deformation in regions with distributed faulting. The block rotation model which is emerging from this approach seems to provide the means for integrating structural, paleomagnetic, geodetic and rock mechanical data so that actual and complex fault patterns insitu may become decipherable. Several important problems and obstacles still remain however in applying the model. One of the most obvious unknowns is the nature of the boundaries which surround the faulted domain. As the model implies, the deformational incompatibility (i.e. fig. 1) which arises at the edges of the rotating blocks must be absorbed at the domain boundary, in the form of gaps, overlaps, distributed deformation, or extensive rock crushing-phenomena which have barely been investigated. It is evident however that the existence of the domain boundary zone, if the model is correct, implies significant strain softening in which the inital rock mass deformation at the boundary enhances further deformation there. The actual mechanics of such softening and the associated kinematics remain to be unravelled.

The discovery that the resolved shear stress on the San Andreas fault (Zoback, et al., 1988) and by inference other major faults is very low may be relevant to the mechanical behavior of domain boundaries, which often consists of such faults. It is possible that it is the strain softening process in which the stiffness of the boundary decreases with increasing deformation, which is responsible for the very low shear stress needed to cause slip on a fault with large offset.

A closely related problem is the nature of the lower boundary of a domain. In McKenzie's "vorticity model" (McKenzie & Jackson, 1983) the rotating domains of brittle crustal rocks are tightly coupled to the underlying creeping or flowing ductile rock mass. In contrast, Freund's block rotation model requires widespread differential motion between the rotating blocks and the underlying crust, in the form of prevasive detachment surfaces. It is unclear how deep these surfaces might be, and how is the differential motion accommodated. It is reasonable however to expect detachment at depths corresponding to the brittle-ductile transition, perhaps on the order of 10 km depth. Whatever the mechanism, the detachment surfaces, according to this model, must accommodate relatively small displacements, mostly in the form of local rotations of the overlying blocks. The common, perhaps ubiquitous horizontal strong seismic reflectors often seen in the mid continental crust are the most likely sites for the required detachment, but a much more systematic study using active seismic imaging coupled with structural analysis in regions with fault domains is required to clarify this issue.

The contrast in mechanical behavior of the upper rotating "rigid" blocks in fault domains, the underlying "ductile" crust, and the strain softening behavior of the rock mass at the domain boundaries highlight the general issue of quantifying the deformational behavior of the rocks involved. As seen in the section on stress-strain relation for a single domain, a single domain exhibits a high degree of mechanical anisotropy--due to the singular direction of the active fault set. Consequently, the full relation between incremental stress and strain is likely to be fairly complicated perhaps like such relation in solids with monoclinic or even triclinic symmetry. Furthermore, this anisotrpic relation will be, as shown earlier, significantly nonlinear, depending on the initial orientation of faults relative to the principal stresses, the magnitude of block rotation, and the coefficient of friction, etc. Such constitutive relations are much more complicated than the simple elastic relation which are commonly invoked for describing crustal deformation such as around active faults and lithospheric flexure. Consequently much better understanding of these constitutive relations in real rock masses is needed.

Domain deformation via block rotation is dependent on the orientation of the faults within the domain. But the origin of these faults is obviously dependent on the tectonic process. The unanswered question here is the origin of the primary active faults within a domain. Although in laboratory fracture studies rocks and brittle solids in general initially develop two conjugate faults, only one of these faults accommodates further slip. The factors which determine which one of the conjugate directions will dominate insitu are totally unclear. It appears that, if the block rotation model is correct, in compressional or extentional tectonic settings both conjugate direction accommodate the deformation, not in the form of separate conjugate faults, but as neighboring domains of conjugate fault sets. However in pure transform settings, especially along oblique subduction zones, it is found that block rotations almost invariably have the same sense as the tectonic shear system. The "ball bearing" model of course predicts exactly this kind of rotation, whereas the block rotation model is consistent only if the fault which accommodate these rotations are oriented appropriately. Unfortunately very little knowledge is available regarding the nature of the faults which accommodate rotations about vertical axes in oblique subduction zones (Beck, 1976). How the process of oblique subduction creates the associates strike slip domains remains another important problem for future research.

Conclusions

The kinematics of block rotations by fault sets in domains, as first proposed by Freund (1970) provide an attractive model for understanding and analyzing crustal deformation when pervasive faulting is involved. The model is especially important because crustal deformation by pervasive faulting may very well be the rule.

Mechanical considerations of rock and fault strength and frictional properties suggest that new fault sets may form when older ones have rotated sufficiently away form the direction of optimal failure. Thus the combination of block and fault rotation with fault mechanics provides a simple model which can explain rather complex fault patterns in situ.

The rotation of blocks and faults due to fault slip in a distributed fault system must be included in the analysis of geodetic and other crustal deformation data, because it can significantly modify the interpretation of such data. Finally, the mechanics of block and fault rotations cause heterogeneities at block and domain boundaries. These heterogeneities may be responsible for the initiation and arrest of rupture on active faults. This is true especially in distributed fault systems, where the prediction of an earthquake involves not only time and magnitude, but also the actual fault on which the event is to occur.

Acknowlegment. This study was supported by the Geodynamics Program of NASA, through grant no. NAG5-926. Comments by Bruce Luyendyk and Peter Bird very much helped clarify several important points. We thank them both.

References

Anderson, E.M., 1951, The dynamics of faulting: *Oliver and Boyd, Edinburgh,* p. 206.

Anderson, R.E., 1973, Large-magnitude late Tertiary strike-slip faulting north of Lake Mead, Nevada: *U.S. Geological Survey Professional Paper 794.*

Anderson, R.E., Longwell, R.C., Armstrong, R.L., and Marvin, R.F., 1972, Significance of K-Ar ages of Tertiary rocks from the Lake Mead region, Nevada-Arizona: *Geological Society of America Bulletin,* v. 83, pp. 273-288.

Angelier, J., Collette, B., and Anderson, E.R., Neogene paleostress changes in the Basin and Range: a case study at Hoover Dam, Nevada-Arizona: *Geological Society of America Bulletin,* v. 96, pp. 347-361.

Aydin, A., and Page, B., 1984, Diverse Pliocene-Quaternary tectonics in a transform environment: San Francisco Bay region, California: *Geological Society of America Bulletin,* v. 95, pp. 1303-1317.

Aydin, A., and Page, B., 1984, Diverse Pliocene-Quaternary tectonics in a transform environment: San Francisco Bay region, California: *Geological Society of America Bulletin,* v. 95, pp. 1303-1317.

Beck, M.E., Jr., 1976, Discordant paleomagnetic pole position as evidence of regional shear in the Western Cordillera of North America. *American J. Science,* v. 276, p. 694-712.

Billings, M.P., 1972, Structural Geology: *Prentice-Hall,* Englewood Cliffs, N.J., 606 p.

Bohannon, R.G., 1979, Strike-slip faults of the Lake Mead region of southern Nevada, in Armentrout, Cole and TerBest, eds., Cenozoic Paleogeography of the Western United States: Pacific Section, *Society of Economic Paleontologists and Mineralogists,* p. 129-139.

Brown, L.L., and Golombek, M.p., 1985, Tectonic rotation within the Rio Grande rift: evidence from paleomagnetic studies: *Journal of Geophysical Research,* v. 90, no. 1, p. 790-802.

Carr, W.J., 1984, Regional structural setting of Yucca Mountain, southwestern Great Basin, Nevada and California: *U.S. Geological Survey Open-File Report 84-854.*

Carter, B.A., 1982, Geology and structural setting of the San Gabriel anorthosite-syenite body and adjacent rocks of the western San Gabriel Mountains, Los Angeles County, California, in *Geologic Excursions in the Transverse Ranges, a Guidebook,* Geological Society of America Cordilleran Section, 78th Annual Meeting.

Dibblee, T.W., Jr., 1961, Evidence of strike-slip movement on northwest-trending faults in the western Mojave Desert, California: *U.S. Geological Survey Professional Paper 424-B,* p. B197-B199.

Dokka, R.K., 1983, Displacements on late Cenozoic strike-slip faults of the central Mojave Desert, California: *Geology,* v. 11, p. 305-308.

Fagin, S.W., and Gose, W.A., 1983, Paleomagnetic data from the Redding section of the Eastern Klamath belt, northern California: *Geology,* v. 11, p. 505-508.

Frei, L., 1980, Junction of two sets of strike slip faults in southeastern Sinai (in Hebrew with English abstract): *M.Sc. thesis,* Hebrew University, Jerusalem.

Freund, R., 1970a, Rotation of strike-slip faults in Sistan, southeastern Iran: *Journal of Geology,* v. 78, p. 188-200.

Freund, R., 1970b, The geometry of faulting in the Galilee: Israel Jouranl Earth Sciences, v. 19, pp. 117-140.

Freund, R. 1971, The Hope fault, a strike slip fault in New Zealand: *New Zealand Geological Survey Bulletin,* v. 86, 49 p.

Freund, R. 1974, Kinematics of transform and transcurrent faults: *Tectonophysics,* v. 21, p. 93-134.

Freund, R., and Merzer, A.M., 1976, The formation of rift valleys and their zigzag fault patterns: *Geology Magazine,* v. 113, no.6, pp. 561-568.

Gans, P.B., and Miller, 1983, Style of mid-Tertiary extension in east-central Nevada, in *Guidebook, pt. 1,* Geological Society of America Rocky Mountain and Cordilleran Sections Meeting, Utah Geology and Mining Survey Special Studies, v. 59, pp. 107-160.

Garfunkel, Z., 1970, The tectonics of the western margins of the southern Arava: *Ph.D. dissertation*, Hebrew University, Jerusalem, 104 p.

Garfunkel, Z., 1974, Model for the late Cenozoic tectonic history of the Mojave Desert, California and for its relation to adjacent areas: *Geological Society of America Bulletin*, v. 85, p. 1931-1944.

Greenhaus, M.R., and Cox, A., 1979, Paleomagnetism of Morro Rock-Islay Hill complex as evidence for crustal block rotation in central coastal California: *Journal of Geophysical Research*, v. 85, no. 5, p. 2393-2400.

Hill, R.L., and Beeby, D.J., 1977, Surface faulting associated with the 5.2 magnitude Galway Lake earthquake of May 31, 1975: Mojave Desert, San Bernadino County, California: *Geological Society of America Bulletin*, v. 88, p. 1378-1384.

Hoeppener, R., Kalthoff, E., and Schrader, P., 1969, Zur physikalischen Tektonik: Bruchbildung bei verschiedenen Deformationen im Experiment: *Geol. Rundsch*, v. 59, p. 179-193.

Hornafius, J.R., 1985, Neogene tectonic rotation of the Santa Ynez Range, western Transverse Ranges, California, suggested by paleomagnetic investigation of the Monterey Formation: *Journal of Geophysical Research*, v. 90, no. 14, p. 12503-12522.

Hudson, M.R., and Geissman, W.J., 1985, Middle Miocene counterclockwise rotation of rocks from west-central Nevada: implication for Basin and Range extension: *Geological Society of America, Abstracts and Program*, Annual Meeting, Orlando, Florida, no. 59705, p. 615.

Hutton, L.K., Johnson, C.E., Pechmann, J.C., Ebel, J.E., Given, T.W., Cole, D.M., and German, P.T., 1979, Epicentral locations for the Homestead Valley earthquake sequences: *California Geology*, v. 33, p. 110-164.

Kamerling, M.J., and Luyendyk, B.P., 1985, Paleomagnetism and Neogene tectonics of the Northern Channel Islands, California: *Journal of Geophysical Research*, v. 90, no. 14, pp. 12485-12502.

Luyendyk, B.P., Kamerling, M.J., and Terres, R., 1980, Geometric model for Neogene crustal rotations in southern California: *Geological Society of America Bulletin*, v. 91, p. 211-217.

MacDonald, D.W., 1980, Net tectonic rotation, apparent tectonic rotation, and the structural tilt correction in paleomagnetic studies: *Journal of Geophysical Research*, v. 85, pp. 3659-3669.

McFadden, B.J., Opdyke, N.D. and Woodburne, M.O., 1987, Paleomagnetism of the Middle Miocene Barstow Formation, Mojave Desert, Southern California: Magnetic polarity, stratigraphy and tectonic rotation, *EOS*, 68, 16, p. 291.

McKenzie, D. and Jackson, J., 1983, The relationship between strain rates, crustal thickening, paleomagnetism, finite strain and fault movements within a deforming zone, *Earth Planetary Science Letters*, 65, pp. 182-202.

Morton, J.L., and Hillhouse, J.W., 1983, Paleomagnetism and K-Ar ages of Miocene basaltic rocks in the western Mojave Desert, California: Unpublished manuscript.

Nadai, 1950, Theory of Fracture and Flow of Solids: *McGraw-Hill, Longon*, v. 1.

Nicholson, C., Seeber, L., Williams, P.L., and Sykes, L.R., 1986, Seismic deformation along the southern San Andreas fault, California: *Journal of Geophysical Research*, in press.

Nur, A., and Helsley, C.E., 1971, Paleomagnetism of Tertiary and recent lavas of Israel: *Earth and Planetary Science Letters*, v. 10, p. 375-379.

Nur, A., Ron, H., and Scotti, O., 1986, Fault mechanics and the kinematics of block rotation: *Geology*, v. 14, p. 746-749.

Proffet, J.M., Jr., 1977, Cenozoic geology of the Yerington district, Nevada, and implication for the nature and origin of Basin and Range faulting: *Geological Society of America Bulletin*, v. 88, p. 247-266.

Rogers, A.M., and Lee, W.H.K., 1976, Seismic study of earthquakes in the Lake Mead, Nevada-Arizona region: *Seismological Society of America Bulletin*, v. 66, no. 5, p. 1631-1657.

Ron, H. Aydin, A., and Nur, A., 1986, The role of strike-slip faulting in the Late Cenozoic deformation of the Basin and Range Province: *Geology,* in press.

Ron, H., Freund, R., Garfunkel, Z., and Nur, A., 1984, Block rotation by strike-slip faulting: Structural and paleomagnetic evidence: *Journal of Geophysical Research*, v. 89, no. B7, pp. 6256-6270.

Sauber, J., Thatcher, W., and Solomon, S.C., 1986, Geodetic measurement of deformation in the Central Mohave Desert, California: *Journal of Geophysical Research*.

Stein, R.S., and Lisowski, M., 1983, The 1979 Homestead Valley earthquake sequence, California: Control of aftershocks and postseismic deformation: *Journal of Geophysical Research*, v. 88, no. B8, pp. 6477-6490.

Terres, R.R., and Luyendyk, P.B., 1985, Neogene tectonic rotation of the San Gabriel region, California, suggested by paleomagnetic vectors: *Journal of Geophysical Research*, v. 90, no. B7, pp. 12,467-12,484.

Terres, R.R., and Sylvester, A.O., 1981, Kinematic analysis of rotated fractures and blocks in simple shear: *Seismological Society of America Bulletin*, v. 71, no. 5, pp. 1593-1605.

Wells, R.E., and Coe, R.S., 1985, Paleomagnetism and geology of Eocene volcanic rocks of southwest Washington, implication for mechanisms of tectonic rotation: *Journal of Geophysical Research*, v. 90, no. 2, pp. 1925-1947.

Young, K.D., Jancin, M., Voight, B., and Orkan, N.I., 1985, Transform deformation of Tertiary rocks along the Tjornes fracture zone, north-central Iceland: *Journal of Geophysical Research*, v. 90, no. 12, pp. 9986-10010.

Zoback, M.L., and Zoback, M.D., 1980, Faulting pattern in north-central Nevada and strength of the crust: *Journal of Geophysical Research*, v. 85, no. B1, pp. 275-284.

Zoback, M., Zoback, M.L., Mount, V.S. et al., 1987, New Evidence on the State of Stress of the San Andreas Fault System, *Science,* v. 238, pp. 1105-1111.

PLATE MOTIONS, EARTH'S GEOID ANOMALIES, AND MANTLE CONVECTION

Fu Rong-shan

Department of Earth and Space Sciences
University of Science and Technology of China

Abstract. A thermal-convection model constrained on an homogeneous, isoviscous, and internally heated mantle is employed to investigate the relationship between plate motions, earth's geoid anomalies, and mantle-convection patterns. Results show that the correlation between the poloidal component of the velocity of plate motions and the surface-velocity field of convection models associated with the earth's geoid anomalies in the degree and order range of 4 to 6 is about 0.6 and has a maximum value at degree 4. The poloidal motions of plates are either parts of the whole mantle convection (the rigid lithosphere just plays the role of thermal-boundary layers of the mantle convection) or they are caused by viscous drag forces (the rigid lithosphere is independent from the mantle convection). Further work is needed to explain the toroidal motions of plates and their poloidal motions with degree and order 1.

Introduction

The correlations between the earth's geoid anomalies and the tectonic features of the earth's surface have been noted by many studies (e.g., Kaula, 1972). Runcorn (1964, 1967) pioneered the idea that geoid anomalies simply reflect the density differences which drive mantle convection and developed the equations which are used to calculate sub-lithospheric stress patterns with low-degree geoid coefficients. Following his work, many studies investigated sub-lithospheric stress patterns up to degree and order 30 of the earth's geoid anomalies (Liu, 1977, 1978, 1980, a, b, Liu et al. 1976; Huang and Fu, 1982) and the stress field in the lithosphere (Fu and Huang, 1983). The convection models that are connected with the surface features of the plate motions had been established by Hager and O'Connell (1979). Unfortunately, the thermal dynamic processes haven't been taken into account in his work. Recently, Fu (1986a, b) developed a thermal-convection model, which is in an homogeneous, isoviscous, and internally heated spherical shell and constrained to fit the earth's geoid anomalies, and investigated the effects of boundary conditions and Rayleigh numbers on convection patterns. Both studies (Hager, 1979, and Fu, 1986a, b) directly focused on looking for the relationship between convection patterns and observed geophysical data. A general discussion of the correlations between the plate motions and the earth's geoid was made by Peltier (1985). He found an increasingly strong negative correlation between the geoid and $\nabla H \cdot u$ from degree 2 to degree 8 and a positive correlation between the geoid and $Z \cdot (\nabla H \times u)$ in the same range. Another source of relevant data is the seismic tomography of Dziewonski (1984), which Hager et al, (1985) found to indicate density variations in agreement with the geoid pattern of a dynamic earth model. Their work shows that the heterogeneity is throughout the entire mantle and is related to mantle convections. More recently, Jarvis and Peltier (1986) tried to make a bridge between the thermal structure in the convection mantle and its lateral heterogeneity, determined by seismic tomography with spectral analysis.

The purpose of this paper is to investigate the correlations between the surface-velocity field of the mantle thermal-convection model suggested by Fu (1986a, b) and the absolute motions of the plates. A general discussion will be made to explain the toroidal component of the plate motions by the coupling between the mantle viscous flow and the rigid lithosphere.

Thermal-convection Model

Consider a spherical shell containing an incompressible fluid and a symmetric gravitational field. In the Boussinesq approximation, the basic equations can be written as follows:

$$\frac{\partial U_i}{\partial t} = -\frac{\partial(\delta p/\delta)}{\partial X_i} + \alpha_1 g(r) X_i \theta + \nu \nabla^2 U_i$$

$$\partial \theta / \partial t = -U_i \partial T / \partial X_i + \kappa \nabla^2 \theta \qquad (1)$$

$$\partial U_i / \partial X_i = 0$$

Copyright 1989 by
International Union of Geodesy and Geophysics
and American Geophysical Union.

where α_1 is the coefficient of volume expansion, δp is the perturbation of pressure, θ is the perturbation of temperature, U_i is the velocity vector, X_i is the position vector, ν is the kinematic viscosity, and κ is the thermal conductivity. The initial distribution of temperature in the shell is

$$T = \beta_0 - \beta_2 r^2 + \beta_1/r \qquad (2)$$

where β_0, β_1 and β_2 are constants and $\beta_2 = e/6$, where e is the volumetric heating.

The gravitational potential V must satisfy Poisson's equation

$$\nabla^2 V = 4\pi G\rho$$

where ρ is density and G is the universal gravitational constant. For steady state, the solutions are assumed to be of the form

$$\begin{aligned} X_i\omega_i &= Z(r)\,Y_\ell^m, \\ X_i U_i &= W(r)\,Y_\ell^m, \\ \theta &= H(r)\,Y_\ell^m, \end{aligned} \qquad (4)$$

and, considering $Z = 0$ where ω_i is the vorticity,

$$Y_\ell^m(\theta,\phi) = \binom{\cos m\phi}{\sin m\phi} P_\ell^m(\cos\theta)$$

where $P_\ell^m(\cos\theta)$ are the associated Legendre polynomial. Chandrasekhar (1961) has rewritten equation (1) in the following form:

$$\begin{aligned} DL\, F &= -L(L+1)\, CL\, W \\ DL\, W &= F \end{aligned} \qquad (5)$$

where

$$DL = \frac{d^2}{dr^2} + \frac{2}{r}\frac{d}{dr} - \frac{L(L+1)}{r^2}$$
$$F = L(L+1) R1^4 \gamma/\nu\, H$$

where $CL = \alpha_1 \beta R1^6/\kappa\nu$ is the Rayleigh number, $R1$ is the radius of the sphere, $\gamma = \alpha_1 g(r)$, and $\beta = \beta_2 + \beta_1/2r$.

Choosing the half-order Bessel function as the basis function, Fu (1986a, b) solved equation (5) as follows

$$W^\ell = [J_{\ell+1/2}(\alpha r) + J_{-(\ell+1/2)}(\alpha r)]/\sqrt{r} \qquad (6)$$

Since the complete solution is the linear superposition of all the basis solutions which satisfied equation (5), then

$$r\, U_r = \sum_{\ell=0}^{\infty} \sum_{j=1}^{6} B_j\, W_j^\ell(\alpha_j r)\, Y_\ell^m(\theta\cdot\phi) \qquad (7)$$

where B_j are constants to be determined by boundary conditions.

By taking Runcorn's equations (1967) as a boundary condition and choosing another five boundary conditions, we can form a group of linear equations to obtain the B_j. That is:

$$A\,B = F \qquad (8)$$

where

$$A = \begin{pmatrix} W_j^\ell(1\alpha_j) \\[4pt] W_j^\ell(\eta\alpha_j) \\[4pt] \alpha_j^4 W_j^\ell(\alpha_j 1) \\[4pt] \alpha_j^4 W_j^\ell(\alpha_j \eta) \\[4pt] \dfrac{d^2 W_j^\ell(\alpha_j 1 \text{ or } \eta)}{dr^2}\ \left[\text{or } \dfrac{dW_j^\ell(\alpha_j 1 \text{ or } \eta)}{dr}\right] \\[8pt] \dfrac{d^2 W_j^\ell(\alpha_j 1)}{dr^2} + 2\dfrac{dW_j^\ell(\alpha_j 1)}{r\,dr} - \eta^{\ell+1} \\[8pt] \left[\dfrac{d^2 W_j^\ell(\alpha_j \zeta\eta)}{dr^2} + \dfrac{2}{r}\dfrac{dW_j^\ell(\alpha_j \eta)}{dr}\right] \end{pmatrix}$$

$$j = 1, 6$$

$$B = (B1, B2, B3, B4, B5, B6)^T,$$
$$F = (0, 0, 0, 0, 0, FL)^T,$$
$$FL = \left[\frac{A1}{A2}\right]^{\ell+1} \frac{2L+1}{L+1} \frac{Mg}{4\pi\mu A2^2},$$

$\eta \triangleq r_c/R1$ and r_c is the radius of the lower boundary, $A2$ is the radius of the lower boundary also, $A1$ is the radius of the upper boundary, M is the mass of the earth and μ is the viscosity of the mantle. Then, the final solution of this problem is

$$W = \sum_\ell \sum_m \sum_j B_j\, W_j^\ell\, (C_\ell^m \cos m\phi + S_\ell^m \sin m\phi) P_\ell^m(\cos\theta) \qquad (9)$$

where C_ℓ^m, S_ℓ^m are the harmonic coefficients of the earth's geoid anomalies. Using equation (9), obtaining the convection field of the mantle as well as the density distributions in the mantle is simple.

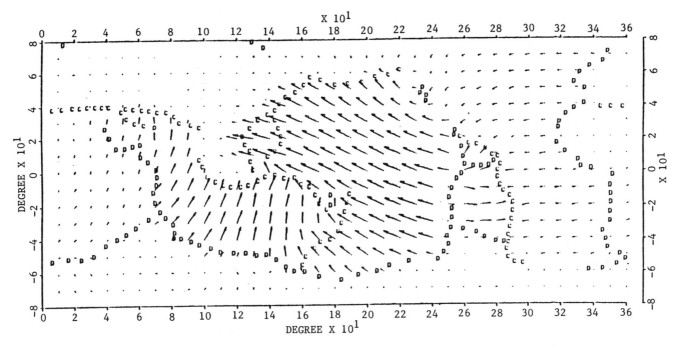

Fig. 1. The plate absolute motions with Model AM1-2 up to degree and order 15. Symbol D is for divergent boundary of plates; Symbol C is for convergent boundary of plates (horizontal axis: longitude from 0°-360° with 20° interval, and vertical axis: latitude from -80°--80° with 20° interval).

Absolute Motions of Plates and Their Harmonic Expansion

Minster and Jordan (1978) obtained a complete analysis of the present-day plate motions, showing that the absolute-motion model AM1-2 of the plates " represents the most satisfactory description available from the present observations ". This model will be used here.

As Hager and O'Connell (1978) and Peltier (1985) had done, the velocity field of the plate motions can be divided into a poloidal field and a

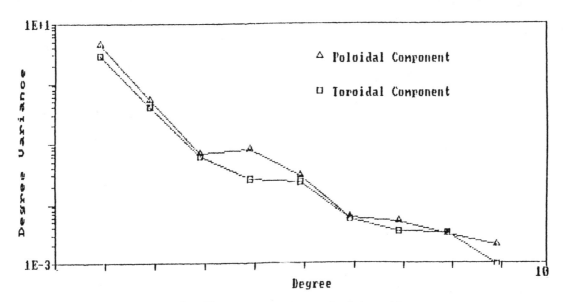

Fig. 2. The power spectrum of plate motions.

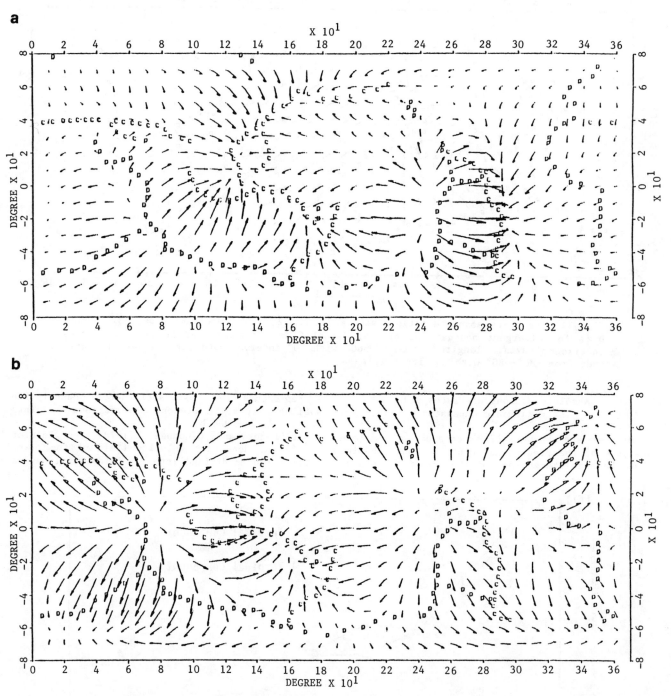

Fig. 3. Velocity field with degree and order 2 to 8; a) poloidal field of the plate motions (AM1-2), b) computed convection pattern (model I); (same as Figure 1).

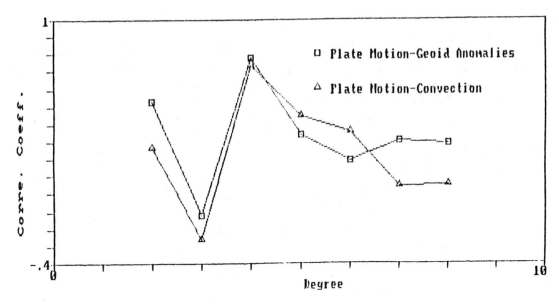

Fig. 4. Correlation between convection and plate motions.

toroidal field. From the plate absolute-motion model AM1-2, the spherical harmonic expansion of the poloidal and toroidal terms up to degree and order 20 are computed with the formula

$$S_\ell^{m\binom{c}{s}} = \frac{1}{4\pi\ell(\ell+1)} [v_\theta \frac{\partial Y_\ell^{m\binom{c}{s}}}{\partial \theta} + \frac{v_\phi}{\sin\theta} \frac{\partial Y_\ell^{m\binom{c}{s}}}{\partial \phi}] \, d\phi \, d\cos\theta$$

$$T_\ell^{m\binom{c}{s}} = \frac{1}{4\pi\ell(\ell+1)} [\frac{v_\theta}{\sin\theta} \frac{\partial Y_\ell^{m\binom{c}{s}}}{\partial \phi} - v_\phi \frac{\partial Y_\ell^{m\binom{c}{s}}}{\partial \theta}] \, d\phi \, d\cos\theta \quad (10)$$

where $T_\ell^{m\binom{c}{s}}$ are the coefficients of the toroidal field, $S_\ell^{m\binom{c}{s}}$ are the coefficients of the spheroidal field, $v_\theta(\theta,\phi)$ and $v_\phi(\theta,\phi)$ are the north-south and east-west components, respectively, of the velocity field of the plate motions. The restructuring of the plate absolute motions up to degree and order 15 are shown in Figure 1, where D and C symbolize the divergent and convergent boundary, respectively. Figure 2 illustrates the power spectrum of plate absolute motions. Both poloidal and toroidal components of the plate motions are nearly the same.

Computation Models and Results

Parameters used in the models throughout the calculations are the earth's radius, RO = 6371 km; core-mantle boundary (CMB) radius, Rc = 3480 km; radius of the mantle-lithosphere boundary (LMB), Rl = 6280 km; and the mantle viscosity, $\mu = 10^{22}$

TABLE 1. Boundary Models and Correlation Coefficients.

MODEL	BOUND.	RAYL. NUM.	2 - 8	4 - 6	2 - 6
MODEL I	CMB FREE	4.5E4	-0.1915	-0.2686	
	LMB FRIC.	1.0E5	0.1699	0.6166	
MODEL II	CMB FRIC. FREE SURFACE	5.0E4	0.1690	0.5740	0.2419
MODEL III	CMB RIGID LMB FRIC.	1.6E5	0.2169	0.6078	

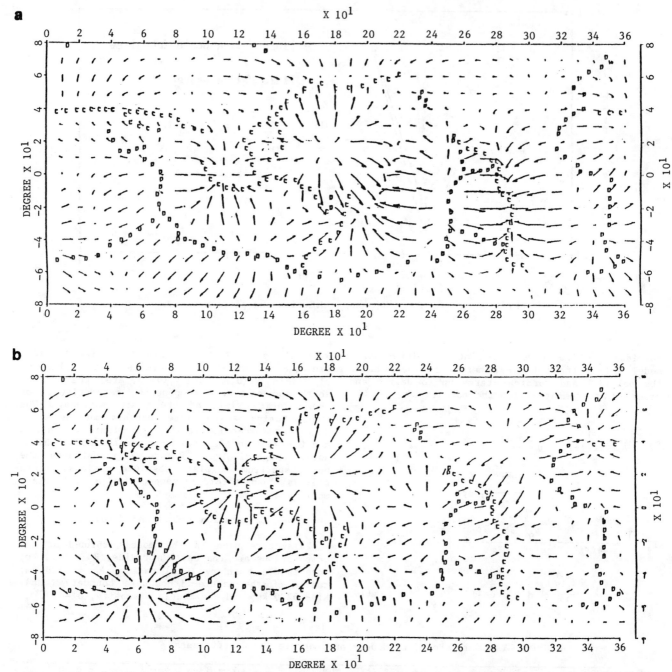

Fig. 5. Velocity field with degree and order 4 to 6; a) poloidal field of the plate motions (AM1-2), b) computed convection pattern (model I); (same as Figure 1).

poise. The geoid model published by Lerch et al. (1979). Three suitable convection models (Table 1) are chosen to compare with the plate motions. Figure 3a is the surface-velocity field of Model I with degree and order 2 to 8. Comparing this model with the poloidal field (Figure 3b) of the plate motions in the same range, we find that these patterns are similar to each other. For example, the motions of the Pacific plate, and the Nazca plate are similar to the calculated values. Clearly, however, the absolute motions of the Europe-Asia plate are quite different from the computed model values. The other major difference between the models and the plate absolute motion is located in Iceland and the North Atlantic, where two patterns appear contradictory.

The correlation coefficient between two fields expanded in harmonics formerly was calculated with the following formula (Kaula, 1980)

$$\rho_\ell(x,y) = \sum_m \{C_\ell^m(x) \cdot C_\ell^m(y) + S_\ell^m(x) \cdot S_\ell^m(y)\}$$

$$/[(2L + 1)\sigma_\ell(x)\sigma_\ell(y)] \quad (11)$$

However, this equation is inconvenient for investigating the correlations between the plate motions and convection patterns, because the power spectrum of the thermal convection is controlled by different factors (e.g., the Rayleigh number and boundary conditions). The basic equation of correlation between two fields $V_c(x,y)$ and $V_p(x,y)$ can be written

$$\rho(v_c, v_p) = \frac{\overline{\sum_{xy}}[V_{cx} \cdot V_{px} + V_{cy} \cdot V_{py}]}{[\overline{\sum_{xy}} V_p^2 \cdot \overline{\sum_{xy}} V_c^2]^{1/2}} \quad (12)$$

The correlation between the poloidal component of plate motions and the thermal-convection pattern as a function of harmonic degree is illustrated in Figure 4. The peak correlation is found at degree 4, with a high correlation of about 0.6 between these two fields in the degree band 4-6 for all three models (Table 1). The maps of convection pattern and plate absolute motions in this range are more compatible (Figure 5).

Discussion and Conclusions

The observation that the earth's geoid anomalies do not directly reflect the density distributions on the earth's surface implies that the earth's geoid anomalies must be associated with the deep-mantle thermal processes (Peltier, 1985). The general agreement between the poloidal field of the plate absolute motions and the thermal-convection patterns obtained from the earth's geoid anomalies in the range degree and order 2 to 8, particularly between degree and orders 4-6, shows a high degree of correlation between these two geodynamical processes. So it is apparent that the deep-mantle thermal processes associated with the earth's geoid anomalies are an important key to explain both observed data.

However, several problems arise before arriving at a plausible answer. First, the harmonic analysis of the plate motions shows that the toroidal and poloidal components have almost the same order in the power spectrum (Hager and O'Connell, 1978; Peltier, 1985). Young (1974) used the nonlinear terms of the momentum equation to explain the toroidal component in the three dimensional thermal convection model and noted that it is quite small in steady cases. A suggestion was made by Hager and O'Connell (1978), in which the effect of the rigid lithosphere is approximated as a strong thermal-boundary layer. Peltier (1985) considered that the continents, which have different chemical compositions from the underlying mantle, probably play a crucial role in developing toroidal motions. In fact, both Hager and Peltier expressed the same idea that the presence of the rigid lithosphere overlying the convecting mantle is important in investigating the nature of the plate absolute motions, including both poloidal and toroidal components with same-order power spectrum. Fu's work (1986a, b) shows that the thermal-convection patterns depend upon the Rayleigh number as well as boundary conditions, and three models (I, II, and III) can fit the earth's geoid anomalies to the plate poloidal motions. However, we can't reject a possibility that much more complicated boundary conditions will dominate thermal-convection patterns. The thickness of the rigid lithosphere is much different from place to place, the temperature distribution is not uniform at the boundary between the lithosphere and the mantle, and the rigid lithosphere is divided into several relative moving parts (plates) by quite different mechanical boundaries. It can be imagined that plate-boundary distributions and physical-chemical conditions do not only manage the patterns of the plate motions but probably also play an important role in the transformations between the poloidal and the toroidal motions.

As an example, throw some wood blocks which have different forms in a small swimming pool with small areas between these blocks. When we heat this pool from the bottom, convection will occur in the water and the wood blocks will move. It is easy to imagine that because of their interactions the motions of wood blocks do not only have poloidal components but also have toroidal components. The second problem is that the degree and order 1 components are presented in the plate absolute motions. What kind of forces drive the motions? Hopefully, a plausible answer to this last question will be found in the future.

In summary, the following conclusions can be made: 1) The absolute motions of the rigid plates are correlated with the mantle thermal convection which is associated with the earth's geoid anomalies. 2) The motions of the rigid plates do not simply and completely reflect the mantle flows below. 3) The plate motions are caused by the coupling between the flowing viscous mantle and the rigid lithosphere. They are not only controlled by the thermal-convection patterns of the mantle but also by the boundary conditions between these rigid plates.

This research was sponsored by the Joint Seismological Science Foundation of China No. (86) 008.

References

Chandrasekhar, U. Hydrodynamic and hydromagnetic stability, Clarendon Press, 652 pp., Oxford, 1961.

Dzienwonski, A. M. Mapping the lower mantle: determination of lateral heterogeneity in P velocity up to degree and order 6, J. Geophys. Res., 89, 5929-5952, 1984.

Fu, R. S., and P. H. Huang. The global stress field in the lithosphere obtained from the satellite gravitational harmonics, Phys. Earth Planet. Inter., 31, 269-276, 1983.

Fu, R. S. A numerical study of the effects of boundary conditions on mantle convection models constrained to fit the low degree geoid coefficients, Phys. Earth. Planet. Inter., 44, 257-263, 1986a.

Fu, R. S. The earth's geoid anomalies and the physical mathematical model of the mantle convection (b), Acta Geophysica Sinica, China, 1986b.

Hager, B. H., and R. J. O'Connell. Subduction zone dip angles and flow driven by plate motion, Tectonophysics, 50, 111-133, 1978.

Hager, B. H., and R. J. O'Connell. Kinematic models of large-scale flow in the earth's mantle. J. Geophys. Res., 84, 1031-1048, 1979.

Hager, B. H., R. W. Clayton, M. A. Richards, R. P. Comer, and A. M. Dzienwonski. Lower mantle heterogeneity, dynamic topography, and the geoid. Nature, 313, 541-545, 1985.

Hager, B. H. Subducted slabs and the geoid constraints on mantle rheology and flow, J. Geophys. Res., 89, 3-6015, 1984.

Huang, P. H., and R. S. Fu. The mantle convection pattern and force source mechanism of recent tectonic movement in China, Phys. Earth Planet. Inter., 28, 261-268, 1982.

Jarvis, G. T., and W. R. Peltier. Heteral heterogeneity in the convection mantle, J. Geophys. Res., 89, 435-451, 1986.

Kaula, W. M. Global gravity and tectonics, in: The Nature of the Earth, E.C. Roberston, ed. McGraw-Hill, New York, N.Y., 385-405, 1972.

Kaula, W. M. Material properties for mantle convection consistent with observed surface fields. J. Geophys. Res., 85, 7031-7044, 1980.

Lerch, F. J., S. M. Klosko, R. E. Laubscher, and C. A. Wager. Gravity model improvement using Geos 3 (GEM 9 and 10). J. Geophys. Res., 84, 3897-3904, 1979.

Liu, H. S. Convection pattern and stress system under the African plate, Phys. Earth Planet. Inter., 15, 60-68, 1977.

Liu, H. S. Mantle convection pattern and subcrustal stress field under Asia, Phys. Earth Planet. Inter., 16, 247-256, 1978.

Liu, H. S. Mantle convection and subcrustal stress under Australia, Mod. Geol., 7, 29-36, 1980a.

Liu, H. S. Mantle convection and subcrustal stress under the United States, Mod. Geol., 7, 81-93, 1980b.

Liu, H. S., E. S. Chang, and G. H. Wyatt. Small-scale mantle convection system and stress field under the Pacific plate, Phys. Earth Planet. Inter., 13, 212-217, 1976.

Minster, J. B., and T. H. Jordar. Present-day plate motions, J. Geophys. Res., 85, 5331-5354, 1978.

Peltier, W. R. Mantle convection and viscoelasticity, Ann. Rev. Fluid Mech., 17, 561-608, 1985.

Runcorn, S. K. Satellite gravity measurements and laminar viscous flow model of the earth's mantle, J. Geophys. Res., 69, 4389-4394, 1964.

Runcorn, S. K. Flow in the mantle inferred from low degree harmonics of the geopotential, Geophys. J. R. Astron. Soc., 14, 375-384, 1967.

Young, R. E. Finite-amplitude thermal convection in a spherical shell, J. Fluid Mech., 63, 695-721, 1974.

ROLE OF EPISODIC CREEP IN GLOBAL MANTLE DEFORMATION

G. Ranalli

Department of Earth Sciences, Carleton University and Ottawa-Carleton Geoscience Centre,
Ottawa, Ontario K1S 5B6

H. H. Schloessin

Department of Geophysics, University of Western Ontario,
London, Ontario N6A 5B7, Canada

Abstract. A unifying theory of mantle deformation over different timescales is not yet available. Existing models focus on limited frequency ranges or specific rheologies. In particular, the phenomenological and microphysical connections between the discrete character of many creep episodes and the overall (time- and space-averaged) pattern of mantle flow have not been explored in full.

It is the contention of this paper that episodic creep (that is, time- and space- dependent creep) plays an important role in the deformation of the mantle. Microphysical processes which can lead to episodic creep include transient rheological response for times shorter than some characteristic time, softening related to pressure and temperature changes and phase transitions (transformation plasticity), dynamic recrystallization, and variations in the concentration of volatiles.

Order-of-magnitude estimates of the relative importance of these processes indicate that transient rheology at times characteristic of postglacial rebound (10 ka) may be relevant in the lower, but not in the upper, mantle. The effects of transformation plasticity are difficult to assess, but are probably limited by the rapid rate of recovery at high temperature. Dynamic recrystallization is of local importance only. Variations in volatile concentration, on the other hand, can possibly affect the local strain rate by orders of magnitude. In all cases, episodic creep results in deformation which is highly heterogeneous when viewed at the scale of the mantle.

Introduction

Inferences on the rheology of the mantle are based on inversion of geophysical data, extrapolation of theoretical and experimental creep equations, and analysis of microstructures of subcrustal materials tectonically transported to the surface. All these procedures are subject to large uncertainties. Surface loading does not permit discrimination between different rheologies and is not generally sensitive to fine rheological structure [see Muller, 1986; Ranalli, 1987 for a discussion]. Extrapolation of creep equations requires assumptions not only on the values of parameters such as grain size, chemical environment, and activation enthalpy in the mantle [Paterson, 1987], but also speculative models of the variations of these parameters with depth, including the effects of phase transitions [Ranalli and Fischer, 1984; Ellsworth et al., 1985].

Despite this far from satisfactory situation, a consensus has emerged that the effective viscosity in the mantle varies from a value of $10^{20} - 10^{21}$ Pa s in the upper mantle to an average of $10^{22} - 10^{23}$ Pa s in the lower mantle. However, the creep mechanisms, the nature of the rheology (Newtonian or non-Newtonian), and the role of transient creep, are all questions still open to debate.

A common assumption is that mantle deformation is steady-state. Yet perhaps the most important characteristic to emerge from recent studies of mantle convection [see e.g. Jarvis, 1984; Olson et al., 1987; Gurnis, 1988] is the time-dependence of mantle flow. Indeed, nonlinear systems such as Rayleigh-Benard convection and other dissipative flows often exhibit intermittent, episodic behavior. This system episodic flow arises from the intrinsic dynamics of nonlinear processes [cf. e.g. Thompson and Stewart, 1986]. In addition, time-dependence can also stem from non-steady constitutive behavior of mantle material, which we term constitutive episodic flow.

In this paper, attention is focused on constitutive episodicity. Transient creep, in the microstructural sense, is only one type of episodic creep. Other factors resulting in time-dependent flow include ambient changes in temperature and pressure, especially when

Copyright 1989 by
International Union of Geodesy and Geophysics
and American Geophysical Union.

TABLE 1. Creep parameters for olivine, used as initial (zero-pressure) values for the estimation of transition time. From compilations by Carter and Kirby [1978], Kirby and Kronenberg [1987] and Ranalli [1987]. Activation volume for transient creep is assumed.

(a) Steady-state creep

log A (MPa^{-n} s^{-1})	n	E (kJ mol^{-1})	V (10^{-6} m^3 mol^{-1})
4.5 ± 0.5	3.5 ± 0.5	525 ± 50	17 ± 4

(b) Transient creep

log C (MPa$^{-n'}$s$^{-m'}$)	n'	m'	E' (kJ mol^{-1})	V' (10^{-6} m^3 mol^{-1})
8 ± 1	2.0 ± 0.5	0.4 ± 0.1	(2/3)E≤E'≤E	(2/3)V≤V'≤V

related to phase transitions (Cottrell creep and transformation plasticity); dynamic recrystallization; and variations in the chemical environment. We present a preliminary, semi-quantitative assessment of the role of these various mechanisms in the global (but spatially and temporally heterogeneous) deformation of the mantle.

Transient Creep

The possibility that loads of characteristic duration of 10 ka, such as in postglacial rebound, sample the transient, and not the steady-state, rheology of the mantle has been considered for some time [cf. e.g. Weertman, 1978], without gaining general favor. Recently, however, it has been revived to explain the discrepancy between the near-constant mantle viscosity inferred from postglacial rebound, and the increase with depth by one or two orders of magnitude inferred from geoidal anomalies and microrheology. Postglacial rebound appears to be compatible with transient lower-mantle rheology [Peltier et al., 1986; Yuen et al., 1986]. Furthermore, the inferred long-term lower mantle viscosity is of the right order of magnitude, i.e. 10^{22} - 10^{23} Pa s [Yuen et al., 1986]. The upper mantle, on the other hand, shows no departures from steady-state rheology.

Are these conclusions justifiable on the basis of the present knowledge of the transient rheology of silicates? Unfortunately, the transient and steady-state rheologies for any one rock type under the same conditions are not known with sufficient accuracy. The general form of the transient creep equation for silicate polycrystals at high temperature and small strain is [Goetze and Brace, 1972; Carter and Kirby, 1978; Carter et al., 1981]

$$\epsilon_T = C\sigma^{n'}t^{m'}\exp(-H'/RT) \quad (1)$$

where ϵ_T is the transient creep strain, σ is stress, t is time, and T is absolute temperature. The factor C, the stress and time exponents n' and m', and the activation enthalpy H' = E' + pV' (where E' and V' are activation energy and volume, respectively, and p is pressure) are characteristics of the material. At larger strains ($\epsilon_T \geq 10^{-2}$) data are best fit by an exponential time law [Post, 1977]. Since the strains associated with postglacial rebound are very small, we use equation (1) to derive the transient creep rate at constant stress and temperature

$$\dot{\epsilon}_T = m'C\sigma^{n'}t^{(m'-1)}\exp(-H'/RT) \quad (2)$$

which can be compared with the steady-state creep equation for silicate polycrystals, usually of power-law type [cf. Kirby and Kronenberg, 1987 for a recent review],

$$\dot{\epsilon}_S = A\sigma^n \exp(-H/RT) \quad (3)$$

where H = E + pV. The time necessary for the equalization of transient and steady-state creep rate, that is, the time below which transient rheology is predominant, can be estimated in principle by setting $\dot{\epsilon}_T = \dot{\epsilon}_S$, that is

$$t = B^{1/(m'-1)} \sigma^{\Delta n/(m'-1)} \exp[-\Delta H/(m'-1)RT] \quad (4)$$

where B = A/m'C, Δn = n - n', ΔH = H - H'.

While the steady-state creep parameters for olivine-rich ultrabasic rocks are reasonably well-known [cf. Kirby and Kronenberg, 1987; Ranalli, 1987 for reviews], the transient creep parameters are subject to wide uncertainties [Carter and Kirby, 1978; Carter et al., 1981]. In metals, transient creep often exhibits the same stress and temperature dependence as steady-state creep. In silicates, this is not always the case: both the stress exponent and the activation enthalpy for transient creep may be less than for steady-state creep (see Table 1). For this reason, we estimate the transition time as a function of ΔH and Δn, rather than for a given value of the parameters. We also assume that the pressure-dependence of activation enthalpy is similar in steady-state and transient creep, i.e. the ratio H'/H at any depth is the same as the ratio at zero pressure.

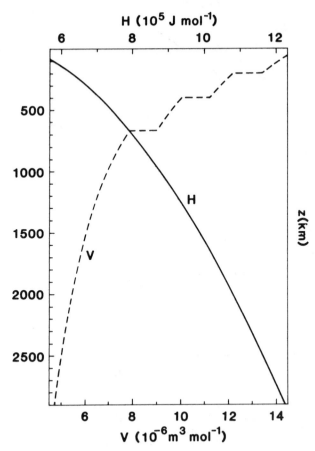

Fig. 1. Activation enthalpy and activation volume for steady-state power-law creep as a function of depth in the mantle [see discussion in Ranalli and Fischer, 1984].

The first-order variations of activation enthalpy and volume in the mantle for steady-state power-law creep are shown in Figure 1. They have been calculated from initial values for a peridotitic upper mantle according to procedures outlined by Sammis et al. [1977] and Ranalli and Fischer [1984], which yield values for steady-state effective viscosities in agreement with geophysical and homologous temperature estimates for the lower mantle [Ellsworth et al., 1985; Heinz and Jeanloz, 1987]. The depth variations of ΔH can be obtained from the experimentally determined ratio at zero pressure. The variations of the parameter B with depth are assumed to be of order one, by analogy with the steady-state creep case. Transition times as a function of ΔH and Δn are shown in Figure 2 for conditions representative of the upper and the lower mantle ($\sigma = 5$ MPa and $T = 1500$ K and 2500 K, respectively). If the characteristic time of postglacial rebound is 10 ka, the condition for the upper mantle to show steady-state rheology is $\Delta H \leq 100$-150 kJ mol^{-1}, while the condition for the lower mantle to show transient rheology is $\Delta H \geq 175$-225 kJ mol^{-1}. This corresponds to an initial (zero-pressure) ratio $H'/H \sim 0.75$-0.85, which is within the experimental range observed in silicates. If this ratio is relatively depth-independent, then transient creep is likely to be predominant in the lower mantle for times characteristic of postglacial rebound, as indicated by geophysical inversion [Yuen et al., 1986]. On the other hand, if the activation parameters in transient and steady-state creep are approximately equal, transient creep is not a significant geodynamic process anywhere in the mantle for times of the order of 10 ka or longer.

The previous considerations are necessarily speculative, as the data for transient creep are uncertain. They do, however, specify the range of conditions for which constitutive transient rheology may be significant in the lower mantle during postglacial rebound.

Cottrell Creep and Transformation Plasticity

The mantle consists of anisotropic grains, and therefore a change in temperature or pressure results in internal stresses that affect the rheology of the aggregate. A phase transformation, with attendant changes in volume and mechanical properties, similarly generates internal stresses.

Thermal cycling can result in a considerable softening of the material [Cottrell, 1964]. The reason for the softness lies in the overcoming of the creep strength by intergranular stresses (Cottrell creep). The same effect can be achieved by variations in hydrostatic pressure, if the polycrystal consists of grains of different compressibilities [Schloessin, 1978]. Pressure-induced stresses (Hertzian stresses) can sometimes lead to disintegration of the material. The local strain associated with intergranular stresses is accommodated by the generation of extra dislocations. It is this increase in average dislocation density that softens the material. At high temperature, when recovery processes are active, the increase in dislocation density due to intergranular stress relaxation and strain accommodation must be faster than the decrease in dislocation density due to recovery for Cottrell creep to contribute significantly to the total strain rate. Temperature (or pressure) cycling must therefore be fast relative to recovery; such fast cycling is difficult to envisage in the mantle.

The softening which is sometimes observed as a polycrystal is cycled through a phase transition (transformation plasticity) can be regarded as a particular case of Cottrell creep. Its macroscopic characterization is sketched in Figure 3. During one cycle of length Δt, the strain rate under constant stress increases, or the flow stress at constant strain rate decreases. Under constant stress, the superplastic strain contributes to the concentration of deformation in the volume where the transition occurs.

Transformation plasticity can be analyzed in a manner analogous to Cottrell creep, by assuming that intergranular stresses are set up by changes in volume and elastic incompatibilities of transforming grains. Poirier [1982] assumed that internal stresses are completely relaxed by extra dislocations, but did not consider recovery. His model has been expanded by

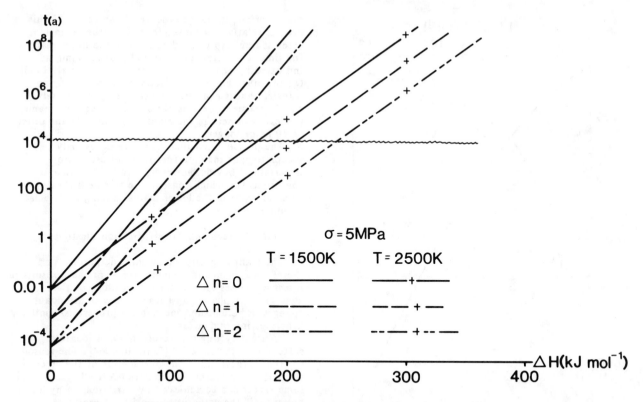

Fig. 2. Transition times (below which transient rheology is predominant) as a function of the differences between steady-state creep and transient creep activation enthalpies and stress exponents. The lower and the higher temperature are representative of the upper and lower mantle, respectively. The wavy line marks the characteristic time for postglacial rebound.

Paterson [1983]. The intergranular stresses generated by the transformation undergo plastic relaxation which increases the dislocation density; this density, however, is also affected by dynamic recovery (see Fig. 3). The total strain occurring during the transition is

$$\Delta\epsilon = \dot{\epsilon}_S \Delta t + \dot{\epsilon}_S t_R (\rho_X/\rho_S) + \epsilon_X \tag{5}$$

where $\dot{\epsilon}_S$ is the steady-state strain rate, ϵ_X the creep strain arising from the transformation, ρ_S and ρ_X the steady-state dislocation density and the density of all the mobile dislocations accommodating the misfit stresses, respectively, t_R the characteristic time of recovery, and Δt the total time of the transition.

The importance of transformation plasticity depends therefore on the relative kinetics of transformation and recovery. When recovery is slow or absent, the contribution of transformation plasticity (besides ϵ_X, which is likely to be small) is of the order of ρ_X/ρ_S; when recovery is fast, the contribution of transformation plasticity is negligible. Paterson [1983] estimated $\rho_X \sim 10^{11}$ m^{-2} for olivine with a grain size of 1 mm (comparable with ρ_S for olivine at 10 MPa). Small grain sizes and low stresses favor transformation plasticity, but the kinetics of transformation and recovery are unknown. At high homologous temperature, recovery should proceed faster than the transformation (the latter being controlled by the rate of change in the ambient parameters, i.e. by slow convective motions). Consequently, it is unlikely that this type of transformation plasticity is important in the mantle.

Plastic relaxation of intergranular stresses is not the only possible mechanism for transformation plasticity. Changes in diffusivities and elastic moduli associated with the transition could change the flow law in the transformed phase. Also, the latter may be relatively dislocation-free, in which case a phase transition would act as a dynamic annealing mechanism. Another possibility has been suggested by Green [1986]. If, under nonhydrostatic stress, the grains of the two phases change shape because they are stable across interfaces of different orientations, a volume-transfer creep results, which may be an important contributor to total strain in the transition zone and does not require cycling of the material through it. However, this depends critically on the possibility of maintaining approximately constant volume fractions of the two polymorphs. It is an interesting possibility which awaits experimental verification.

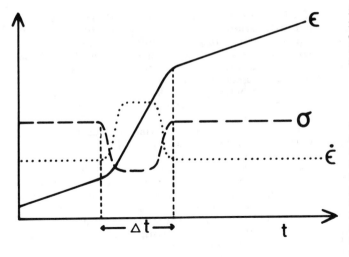

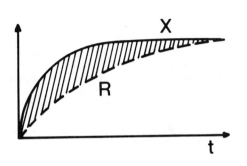

Fig. 3. (Top) Phenomenological characterization of transformation plasticity. During the transformation time Δt, the stress decreases at constant strain rate, or the strain rate increases at constant stress, resulting in a larger strain in the affected volume.
(Bottom) Difference between the kinetics of the transformation (X) and the kinetics of recovery (R): in the case illustrated, as recovery is slower than the transformation, an extra density of dislocations is available (proportional to the shaded area) that contributes to the total strain rate.

Another mode of softening has been suggested by Vaughan and Coe [1981] on the basis of experiments on the olivine → spinel germanate transition, which results in a grain size reduction by about one order of magnitude. This is accompanied by a change from power-law to diffusion creep. However, it should be noted that the experiments were carried out at high stress (~100 MPa), and the kinetics of grain growth are unknown.

In summary, Cottrell creep in the mantle, in the sense of transient enhanced creep related to plastic relaxation of internal stresses (caused by changes in ambient parameters and/or phase transitions), is limited by the relatively fast rate of recovery at high temperature. For transformation plasticity to be important [Sammis and Dein, 1974; Parmentier, 1981], a mechanism which is not dependent on relatively rapid cycling must be operative. Further studies of the kinetics of grain growth and of volume-transfer creep are necessary.

Dynamic Recrystallization

Dynamic recrystallization is common in polycrystals deforming by dislocation creep at high temperature [see review by Poirier, 1985], and probably plays a major role in the development of shear zones and related tectonites (highly deformed, recrystallized rocks) in the upper lithosphere [Urai et al., 1986]. Recrystallized grains may be formed by progressive subgrain rotation or by nucleation and growth: in either case, the macroscopic result is a softening of the material, as strain hardening is reduced. The resulting transient enhancement of the creep rate may lead to the development of plastic instabilities.

Laboratory experiments on silicates at high homologous temperature in which dynamic recrystallization is observed are conducted at relatively high stress (≥ 50 MPa). In metals, dynamic recrystallization occurs only between a low and a high stress cutoff, falling entirely within the dislocation creep field [Sellars, 1978]. There is evidence for both metals and silicates that dislocation density and subgrain size become less sensitive to stress at low stresses; for olivine, the threshold stress is of the order of 10 MPa [Twiss, 1986]. If it is assumed that recrystallized grain size follows a similar relation, dynamic recrystallization in the mantle may be of local importance only, where the stresses are sufficiently high. Bands of recrystallized material are likely to be mainly subhorizontal in the upper mantle, since the general direction of flow is subhorizontal except at ascending and descending plumes. These mantle tectonites are sites of enhanced strain rate and may be related to the subhorizontal seismic reflectors that are present in the upper mantle [Fuchs, 1986].

Besides the transient softening effect caused by syndeformational annealing, dynamic recrystallization under sufficiently large stress is an effective way of reducing grain size. The grain size d is related to the stress by a relation of the type [cf. e.g. Poirier, 1985]

$$d/b = K (\sigma/\mu)^{-r} \qquad (6)$$

where b and μ are Burgers vector and rigidity, respectively, and K is a proportionality factor. The exponent r is usually equal to, or slightly larger than, unity, according to the recrystallization mechanism. If an initially coarse-grained material is subject to a large stress, therefore, the grain size may be sufficiently reduced for the flow mechanism to change from power-law creep to diffusion creep, which is predominant at small grain sizes, of the order of 0.1-1 mm for olivine [see the deformation maps of Frost and Ashby, 1982; Ranalli, 1982].

The likelihood of the occurrence of recrystallized diffusion creep can be assessed by plotting the experimentally determined recrystallized grain size curve

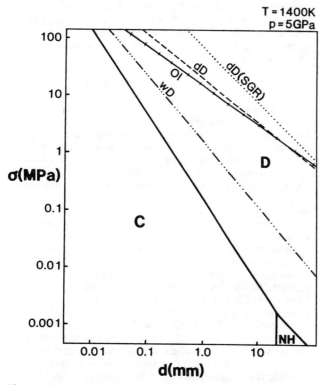

Fig. 4. Stress-grain size deformation map for olivine under upper mantle conditions. C, NH, and D represent the fields of Coble, Nabarro-Herring, and dislocation (power-law) creep, respectively. Recrystallized grain size curves: dD(SGR)-dry dunite by subgrain rotation; dD-dry dunite; Ol-olivine; wD-wet dunite (the last three by nucleation and growth). Data from various sources, discussed in Ranalli [1987].

(equation (6)) on a suitable deformation map [Poirier, 1985; Karato et al., 1986]. Some selected recrystallized grain size curves for ultrabasic silicates are shown in Figure 4, superimposed on a stress-grain size deformation map for olivine under suitable upper mantle conditions [Ranalli, 1982]. All of them fall entirely within the power-law creep field. Stresses larger than 100 MPa and extremely small grain sizes are necessary for dynamic recrystallization to cause a change in the creep mechanism. This conclusion confirms the observations of Zeuch [1983], who found no experimental evidence - either in microstructure or in activation parameters - for a transition to diffusion creep during dynamic recrystallization of olivine. The main effect of dynamic recrystallization in the mantle is likely to be local and limited to discrete shear bands where the stress is sufficiently high, and to consist in removal of strain hardening that results in transient softening while recrystallization occurs.

Volatiles and Chemical Environment

Volatiles (mainly CO_2 and H_2O) are present in the mantle, as evidenced by volcanic emanations. Although volatile release occurs mainly in the outermost shell of the Earth, variations in volatile concentration are likely to be associated with phase transitions, since solubilities change with structure transformations. The distribution of volatiles in the mantle is likely to be highly heterogeneous both in space and time. The geochemical case for such heterogeneities has been made, among others, by Bailey [1980], who argued that they extend to regional scales.

The rheology of natural olivine depends on volatile partial pressures and oxide activities [Ricoult and Kohlstedt, 1985; Karato et al., 1986]. The activation enthalpy of power-law creep is as much as 20-25% less in "wet" olivine than in "dry" olivine. Although it is an oversimplification to discuss the role of volatiles only in terms of this enthalpy effect, it is likely that the overall result of volatile concentration is a decrease in effective viscosity. On this basis, Jackson and Pollack [1987] argued that the average mantle viscosity has increased over time due to the progressive devolatilization of the mantle. Bailey [1980] postulated that, when a plate is fractured and thus offers a conduit of escape for volatiles, it leads to episodic degassing of the mantle beneath - a fact that in itself would cause variations in volatile concentration.

Although the details of the spatial and temporal variations of volatiles in the mantle are impossible to determine, the effect of such variations can be roughly estimated by considering the changes in creep activation enthalpy. If $\dot{\epsilon}_S$ and $\dot{\epsilon}_V$ are the creep rates in a volatile-free and a volatile-rich environment, with activation enthalpies H and H_V, respectively ($H_V = cH$, $c \leq 1$), their ratio is

$$\dot{\epsilon}_V/\dot{\epsilon}_S = \exp[(1-c)H/RT] \qquad (7)$$

where the assumption has been made that changes in the other creep parameters are negligible compared with changes in activation enthalpy. This assumption seems to be valid for olivine, at least as a first approximation [cf. e.g. Karato et al., 1986]. The values of c are usually in the range 0.70-0.90 [see summary by Jackson and Pollack, 1987]. Figure 5 shows the predicted ratio of strain rates as a function of H at a representative mantle temperature. For $500 \leq H \leq 1000$ kJ mol^{-1}, which represents the likely range of activation enthalpies in the mantle, even moderate decreases in activation enthalpy lead to order-of-magnitude increases in creep rate.

If the above arguments are correct (and it must be remembered that the effects of volatile concentration at high pressure are not known), then volatile softening can have important effects in mantle flow. If for instance a relatively volatile-rich plume ascends in a volatile-depleted upper mantle, its lower effective viscosity contributes to its vertical velocity. Thus, in addition to any thermal effect [see Olson et al., 1987], the ascent velocity of plumes can be increased also by their volatile content. Furthermore, temporal and spatial variations in volatile concentration within plumes can contribute to episodicity in hotspot volcanism.

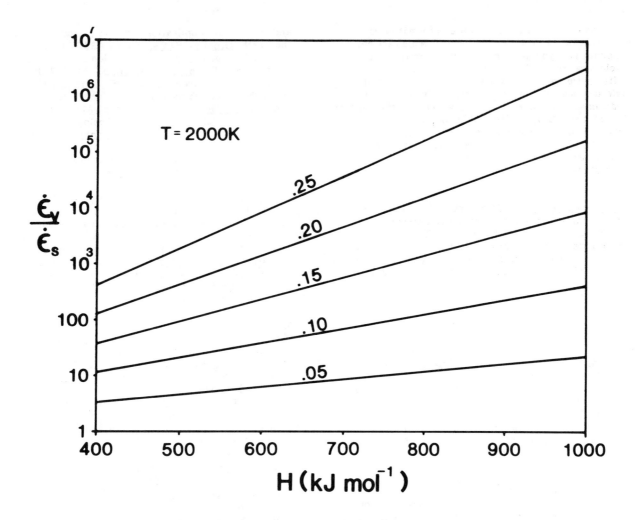

Fig. 5. Variations with activation enthalpy of the ratio of strain rate in the presence of volatiles to strain rate in dry olivine. Numbers on the curves denote the relative decrease of enthalpy (1-c).

Conclusions

Episodicity is an important feature of mantle flow. The sources of non-steady flow can be either systemic (related to the nonlinearity of the governing equations) or constitutive (related to the microphysics of the process). Most current models, therefore, are space- and time-averages that neglect the heterogeneity and intermittency of mantle deformation.

Several factors can generate constitutive episodic creep in the mantle. Their importance and order-of-magnitude effects are as follows:

(1) Transient creep on timescales characteristic of postglacial rebound may occur in the lower mantle if its activation enthalpy is 75-85% of the steady-state creep activation enthalpy. It is unlikely anywhere in the mantle if the two enthalpies are approximately equal.

(2) Cottrell creep and transformation plasticity caused by the plastic relaxation of internal stresses are unlikely to be a major factor, since at high homologous temperature recovery is fast compared with changes in ambient parameters. The likelihood of other softening mechanisms at the phase transition zone (volume-transfer creep) cannot be quantitatively assessed, as the kinetics of grain growth are not known.

(3) Dynamic recrystallization is probably limited to mantle tectonites (i.e. shear bands of stress concentration) which could in turn be related to upper mantle seismic reflectors. It cannot cause a change in creep mechanism from dislocation to diffusion creep for stresses at geodynamic levels.

(4) Volatiles and the chemical environment can change the effective viscosity by a few orders of magnitude as a consequence of their effect on the activation enthalpy. This effect is likely to have important consequences for the thermal history of the Earth, the pattern of convection, and volcanism.

Although the above considerations are only

preliminary, it is clear that mantle deformation is heterogeneous in both time and space. This fact has important geodynamic consequences. The lifetime of a geochemical reservoir, for instance, will be a function of its position and age. Conclusions on the general pattern of mantle convection based on geochemical evidence are consequently not very well constrained [see also the critique by Davies, 1984]. The mantle is neither a nearly-homogeneous fluid in steady motion, nor a stratified fluid. More irregular patterns, with many superpositions of scales and characteristic times, are probably the rule.

Acknowledgements. This work has been supported by the Natural Sciences and Engineering Research Council of Canada. Thanks are expressed to Elsie Lambton, Linda Tapp, and especially to Emergency B. Power for word-processing, and to Harold Ellwand for drafting the figures. The comments of two reviewers have been most helpful.

References

Bailey, D.K., Volcanism, Earth degassing and replenished lithosphere mantle, Phil. Trans. R. Soc. Lond., A 297, 309-322, 1980.
Carter, N.L. and S.H. Kirby, Transient creep and semi-brittle behavior of crystalline rocks, Pure Appl. Geophys., 116, 807-839, 1978.
Carter, N.L., D.A. Anderson, F.D. Hansen, and R.L. Kranz, Creep and creep rupture of granitic rocks, Am. Geophys. Union, Geophysical Monograph, 24, 61-82, 1981.
Cottrell, A.H., The Mechanical Properties of Matter, Wiley, New York, 1964.
Davies, G.F., Geophysical and isotopic constraints on mantle convection: an interim synthesis, J. Geophys. Res., 89, 6017-6040, 1984.
Ellsworth, K., G. Schubert, and C.G. Sammis, Viscosity profile of the lower mantle, Geophys. J. R. Astr. Soc., 83, 199-213, 1985.
Frost, H.J. and M.F. Ashby, Deformation-Mechanism Maps, Pergamon Press, Oxford, 1982.
Fuchs, K., Reflections from the subcrustal lithosphere, Am. Geophys. Union, Geodynamics Series, 14, 67-76, 1986.
Goetze, C. and W.F. Brace, Laboratory observations of high-temperature rheology of rocks, Tectonophysics, 13, 583-600, 1972.
Green, H.W. II, Phase transformation under stress and volume transfer creep, Am. Geophys. Union, Geophysical Monograph, 36, 201-211, 1986.
Gurnis, M., Large-scale mantle convection and the aggregation and dispersal of supercontinents, Nature, 332, 695-699, 1988.
Heinz, D.L. and R. Jeanloz, Measurement of the melting curve of $Mg_{0.9}Fe_{0.1}SiO_3$ at lower mantle conditions and its geophysical implications, J. Geophys. Res., 92, 11437-11444, 1987.
Jackson, M.J. and H.N. Pollack, Mantle devolatilization and convection: implications for the thermal history of the Earth, Geophys. Res. Letts., 14, 737-740, 1987.
Jarvis, G.T., Time-dependent convection in the Earth's mantle, Phys. Earth Planet. Inter., 36, 305-327, 1984.
Karato, S., M.S. Paterson, and J.D. FitzGerald, Rheology of synthetic olivine aggregates: influence of grain size and water, J. Geophys. Res., 91, 8151-8176, 1986.
Kirby, S.H. and A.K. Kronenberg, Rheology of the lithosphere: selected topics, Rev. Geophys., 25, 1219-1244, 1987.
Muller, G., Generalized Maxwell bodies and estimates of mantle viscosity, Geophys. J. R. Astr. Soc. 87, 1113-1141, 1986.
Olson, P., G. Schubert, and C. Anderson, Plume formation in the D"-layer and the roughness of the core-mantle boundary, Nature, 327, 409-413, 1987.
Parmentier, E.M., A possible mantle instability due to superplastic deformation associated with phase transitions, Geophys. Res. Letts., 8, 143-146, 1981.
Paterson, M.S., Creep in transforming polycrystalline materials, Mech. Materials, 2, 103-109, 1983.
Paterson, M.S., Problems in the extrapolation of laboratory rheological data, Tectonophysics, 133, 33-43, 1987.
Peltier, W.R., R.A. Drummond, and A.M. Tushingham, Postglacial rebound and transient lower mantle rheology, Geophys. J. R. Astr. Soc., 87, 79-116, 1986.
Poirier, J.P., On transformation plasticity, J. Geophys. Res., 87, 6791-6797, 1982.
Poirier, J.P., Creep of Crystals, Cambridge University Press, Cambridge, 1985.
Post, R.L., High-temperature creep of Mt. Burnet dunite, Tectonophysics, 42, 75-110, 1977.
Ranalli, G., Deformation maps in grain size-stress space as a tool to investigate mantle rheology, Phys. Earth Planet. Inter., 29, 42-50, 1982.
Ranalli, G., Rheology of the Earth, Allen & Unwin, Boston, 1987.
Ranalli, G. and B. Fischer, Diffusion creep, dislocation creep, and mantle rheology, Phys. Earth Planet. Inter., 34, 77-84, 1984.
Ricoult, D.L. and D.L. Kohlstedt, Experimental evidence for the effect of chemical environment upon the creep rate of olivine, Am. Geophys. Union, Geophysical Monograph, 31, 171-184, 1985.
Sammis, C.G. and J.L. Dein, On the possibility of transformational superplasticity in the Earth's mantle, J. Geophys. Res., 79, 2961-2965, 1974.
Sammis, C.G., J.C. Smith, G. Schubert, and D.A. Yuen, Viscosity-depth profile of the Earth's mantle: effects of polymorphic phase transitions, J. Geophys. Res., 82, 3747-3761, 1977.
Schloessin, H.H., Stresses and spreading resistance at contacts between spheres under high pressure, Phys. Earth Planet. Inter., 17, 22-30, 1978.
Sellars, C.M., Recrystallization of metals during hot deformation, Phil. Trans. R. Soc. Lond., A 288, 147-158, 1978.
Thompson, J.M.T. and H.B. Stewart, Nonlinear Dynamics and Chaos, Wiley, Chichester, 1986.
Twiss, R.J., Variable sensitivity piezometric equations for dislocation density and subgrain diameter and

their relevance to olivine and quartz, Am. Geophys. Union, Geophysical Monograph, 36, 247-261, 1986.

Urai, J.L., W.D. Means, and G.S. Lister, Dynamic recrystallization of minerals, Am. Geophs. Union, Geophysical Monograph, 36, 161-199, 1986.

Vaughan, P.J. and R.S. Coe, Creep mechanism in Mg_2GeO_4: effects of a phase transition, J. Geophys. Res., 86, 389-404, 1981.

Weertman, J., Creep laws for the mantle of the Earth, Phil. Trans. R. Soc. Lond., A 288, 9-26, 1978.

Yuen, D.A., R.C.A. Sabadini, P. Gasperini, and E. Boschi, On transient rheology and glacial isostasy, J. Geophys. Res., 91, 11420-11438, 1986.

Zeuch, D.H., On the inter-relationship between grain size sensitive creep and dynamic recrystallization of olivine, Tectonophysics, 93, 151-168, 1983.

LAYERED BLOCK MODEL IN PROBLEMS OF SLOW DEFORMATIONS OF THE LITHOSPHERE AND OF EARTHQUAKE ENGINEERING

A. D. Gvishiani, V. A. Gurvich, A. G. Tumarkin

Institute of Physics of the Earth, USSR Academy of Sciences,
Moscow, 123810, B. Gruzinskaya, 10

Introduction

A system of interacting bodies partly consisting of absolutely solid ones (rigid blocks) and others - elastic (deformable layers) is considered. The last ones may be non-homogenous and non-linear, i.e. their elastic properties may differ from point to point and Hooke law is not supposed to be fulfilled. We assume that rigid and deformable blocks alternate and their contiguous surfaces remain intact. Thus such a system turns out to be finite-dimensional since its state is completely determined by the position of rigid blocks. It has 3n degrees of freedom in the plane case and 6n degrees of freedom in the spatial one, where n is the number of rigid blocks.

We show that the problem of determination of the state of equilibrium of such system is a problem of convex programming on condition that energy of deformation of an arbitrary layer is a convex function of corresponding generalised coordinates. When deformable blocks obey the Hooke law we obtain a problem of quadratic programming which may be reduced to a system of linear equations.

The case when deformable blocks are thin quadrangular layers separating rigid blocks is of a special interest. Explicit formulae for energy of their deformations and natural conditions for convexity of corresponding functions were obtained in [Gvishiani and Gurvich, 1988].

The proposed model will be referred to as Layered Block Model (LBM). LBM may be applied to problems of dynamics of the lithosphere where thin layers are the fault zones between lithospheric blocks supposed to be rigid. Engineering constructions (primarily buildings) can be also described by LBM considering structural (or temperature) joints between monolithic blocks as thin deformable layers.

Quasistatic processes are predominant in dynamics of lithosphere since masses and moments of inertia of its blocks are great and external forces are slowly changing in time. Thus we may consider the corresponding LBM to be in the state of stable equilibrium and apply to it our results giving the opportunity of its calculation by convex programming methods.

The proposed method gives in principle an opportunity of obtaining time and spatial distribution of deformations, stresses and elastic energy. The knowledge of this distribution enables one to predict places and moments of ruptures, i.e. helps to solve the problem of prediction of earthquakes. However the realization of this program depends on the information on elastic properties and thresholds of tensile strength of deformable layers. The lack of such complete information makes it possible to give only qualitative description of the behaviour of LBM.

The study of dynamics of LBM can be reduced to the problem of solving a system of ordinary differential equations. In this case we drop the convexity assumption and consider an arbitrary LBM. Thus this approach to the problem of calculation of reaction of engineering constructions to dynamic external excitations helps to predict block stuctures response during earthquakes.

Finally let us note that the idea of modelling of quasistatic processes in

Copyright 1989 by
International Union of Geodesy and Geophysics
and American Geophysical Union.

non-homogenuos media by distinguishing absolutely rigid parts (blocks) is not new. This approach was proposed for geomechanical problems in [Cundall, 1971, 1976], and for modelling of the seismic process in [Gabrielov et al., 1986]. Our main result is the possibility of applying to the problem the advanced machinery of convex programming including theoretical results as well as computational algorithms [Rockafellar, 1970, 1967].

Description of the Model

Let us examine a two-dimensional LBM consisting of the set $V=\{v_1,...,v_n\}$ of n rigid blocks and the set $E=\{e_1,...,e_m\}$ of m deformable blocks (layers) (Fig.1).

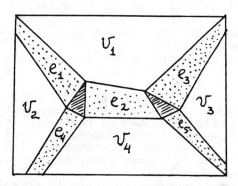

Fig. 1. System of rigid (v) and deformable (e) blocks.

The spatial case is treated completely analogously [Gvishiani and Gurvich, 1988]. We assume that rigid blocks have no contacts and deformable blocks are thin quadrangles.

Let us introduce the following four groups of variables:
1) $x_v=(x^1_v,x^2_v,x^3_v)$ - generalized displacement of the rigid block v from V
   (x^1_v,x^2_v) - displacement of the block's center
   x^3_v - angle of block's rotation
2) $-x_v^*=(-x^{*1}_v,-x^{*2}_v,-x^{*3}_v)$ - generalized external forces applied to v
   $(-x^{*1}_v,-x^{*2}_v)$ - resulting external force
   $-x^{*3}_v$ - resulting external moment
3) $y_e=(y^1_e,y^2_e,y^3_e)$ - generalized deformation of the thin layer e from E
   y^1_e - right (left) shift
   y^2_e - expansion (contraction)
   y^3_e - "cloche" (see Fig.2)
4) $y^*_e=(y^{*1}_e,y^{*2}_e,y^{*3}_e)$ - generalized internal force applied to e

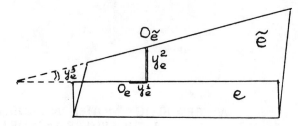

Fig. 2. Generalized deformation of the thin layer e.

y^{*1}_e - shear force
y^{*2}_e - normal force
y^{*3}_e - additional moment.

Now we see that the state of LBM is completely determined by the following 3n-dimensional vectors $x=\{x_{v1},...,x_{vn}\}$, $x^*=\{x^*_{v1},...,x^*_{vn}\}$ and 3m-dimensional vectors $y=\{y_{e1},...,y_{em}\}$, $y^*=\{y^*_{e1},...,y^*_{em}\}$.

Main Theory for LBM

In the case of small deformations x and y are connected by linear operator A (3n·3m matrix) completely determined by the geometry of LBM: $y=Ax$; while y^* and x^* are connected by the adjoint operator A^* (with transposed 3m·3n matrix): $x^*=A^*y^*$.

Total energy of elastic deformation F(y) being the sum of energies of deformation of all layers from E is convex providing F_e is convex with respect to y_e for each e. In [Gvishiani and Gurvich, 1988] it was proved for the case of thin quadrangular layers on condition that local energy of deformation in each point is a convex function of shift and expansion (i.e. if local elastic force is a monotone function of its arguments).

It is also assumed that the potential energy of the system $-G(x)$ in the field of external forces is convex function of generalized displacement x. Then for generalized external forces $-x^*$ we have $-x^*=-dG(x)$, thus x^* is the subdifferential of the concave function $G(x)$. A brief discussion of used facts from convex analysis may be found in Appendix.

The quasistatic case is characterised by the following relation: $x^*_v=x^{*0}_v(t)$ and $x_v=x^0_v(t)$, i.e. the position of rigid blocks and forces acting on them are slowly changing in time. In general the following two cases (the so-called "boundary conditions") are typical:
Dirichlet problem: some blocks are fixed:
$\quad x_v=g^{-1}_v(x^*_v)=x^0_v$,

and others are free from the action of external forces:
$$x^*_v = g_v(x_v) = 0.$$
Neumann problem: each rigid block v from V is under the action of constant external force x^{*0}_v non-depending on x_v with zero resulting force:
$$x^*_v = g_v(x_v) = x^{*0}_v \text{ and } \sum_v x^{*0}_v = 0.$$

All these relations between LBM variables can be described by the following diagram:

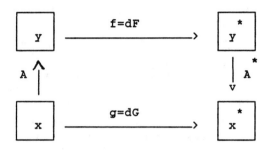

Thus the problem of statics of LBM can be included in the general scheme described by the Rockafellar-Gurvich diagram (see Appendix). Now we are able to apply to it all results from the convex programming presented in the Appendix.

First we observe that P1 for LBM can be rewritten in the following way.

P1 : Find $\bar{x}$ - the equilibrium state for LBM satisfying
$$(A^* dFA - dG)x = 0.$$

From the Main Theorem we get the existence of the solution and the equivalence of P1 to the following problem

P3 : Find $\bar{x}$ minimizing the convex function $-L(x)$ (total system's energy):
$$-L(x) = F(Ax) - G(x) \to \min.$$

The last problem is nothing else but the Lagrange equilibrium principle for LBM.

Results from the Appendix make possible to choose the best way (one of four problems P1-P4) to solve the initial problem in each concrete case.

Dynamics of LBM

Let us now drop the assumption of convexity and consider the behaviour of LBM under a dynamic loading $x^*(t)$. Then as in the static case dynamics of LBM can be described by the following diagram :

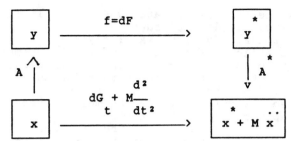

Here A^*x^* - resulting internal force applied to rigid blocks, $x^*(t) = dG_t(x)$ - external forces applied to rigid blocks (varying in time) and $M\ddot{x}(t)$ - inertial forces, M - diagonal matrix with elements equal to masses and moments of inertia of rigid blocks.

Newton law for rigid blocks
$$M\ddot{x}(t) = A^*x^* - x^*$$
makes this diagram closed and gives the main equation of LBM dynamics:
$$M\ddot{x}(t) - (A^* dFA)x = -dG_t(x).$$

Analysis of these equations performed in [Smagin and Tumarkin, 1988] allows to get upper and lower boundaries of maximum possible reaction of LBM to dynamic excitations in terms of their integral characteristics.

Conclusion

Non-linear model of rigid and deformable blocks is established. For the study of statics of LBM four equivalent problems of convex programming are formulated. This makes possible to choose the most convenient method in each concrete case.

Dynamics of LBM can be described by a system of ordinary differential equations which enables us to obtain upper and lower boundaries of system's reaction.

Finally let us note that various problems of physics and economy may be treated on the basis of unified convex programming approach: computation of schemes (electric, hydraulic and gas), optimisation of network flows (transportation pipelines) etc. (Gvishiani and Gurvich, 1984; Gurvich, 1984).

Appendix

Some Facts From Convex Analysis

We shall consider convex and concave functions defined on finite-dimensional spaces and taking real values and the values $\pm\infty$ ([Rockafellar, 1970, §4]). A function F is a proper convex function on R^n if it is not identically $+\infty$ and
$$F(ty + (1-t)y_0) \leq tF(y) + (1-t)F(y_0),$$

whenever $y, y_0 \in R^m$ and $0 < t < 1$. It is closed if its epigraph
$$epi(F) = \{(y,t) \in R^m \times R^1 : f(y) \le t\}$$
is a closed set. We consider only proper and closed convex (concave) functions.

Let $F(y)$ be a convex function on R^m. Its subdifferential operator $f=dF$ is defined to be the multivalued mapping $f: R^m \to R^m$ assigning to each vector $y \in R^m$ the subdifferential $dF(y)$, i.e. the set of its subgradients y^* ([Rockafellar, 1970, §23]): $dF(y) = \{y^* \in R^m : \langle w-y, y^* \rangle \le F(w)-F(y)$ for all $y \in R^m\}$ where $\langle, \rangle$ is denoted the inner product. The operator $g=dG: R^n \to R^n$ is introduced similary for a concave function $G(x)$ on R^n. An exhaustive charactrization of subdifferential operators of convex (concave) functions was obtained in [Rockafellar, 1970, §24].

The conjugates or Young-Fenchel transforms of convex and concave functions are defined by the respective formulas
$$F^*(y^*) = \sup_y [\langle y, y^* \rangle - F(y)], \quad y, y^* \in R^m$$
$$G^*(x^*) = \inf_x [\langle x, x^* \rangle - G(x)] = -(-G)^*(-x^*), \quad x, x^* \in R^n$$

It follows from the Fenchel-Moreau theorem ([Rockafellar, 1970, Theorem 12.2]) that conjugation is an involution on the classes of convex and concave functions, i.e., F^* is convex and $F^{**}=F$, and G^* is concave and $G^{**}=G$.

The subdifferential operators of a pair of conjugate functions are mutually inverse multivalued mappings ([Rockafellar, 1970, Corollary 23.5.1]), i.e., $f=dF \Leftrightarrow f^{-1}=dF^*$, $g=dG \Leftrightarrow g^{-1}=dG^*$.

The effective sets of convex and concave functions are defined in [Rockafellar, 1970, §4] by the formulas
$$dom\, F = \{y \in R^m : F(y) < +\infty\}$$
$$dom\, G = \{x \in R^n : G(x) > -\infty\}.$$

Let C be a convex set in R^m; its relative interior $ri\, C$ is the interior (int) relative to its linear span.

Main Theorem for Rockafellar-Gurvich Diagram

Consider the following diagram :

```
              f=dF
       ┌───┐ ─────────> ┌───┐
       │ y │ <───────── │ y*│
       └───┘   f⁻¹=dF*  └───┘
         ∧                │
       A │                │ A*
         │                v
       ┌───┐    g=dG    ┌───┐
       │ x │ ─────────> │ x*│
       └───┘ <───────── └───┘
              g⁻¹=dG*
```

where $F(y)$ and $F^*(y^*)$ are conjugate convex functions on R^m, $G(x)$ and $G^*(x^*)$ are conjugate concave functions on R^n; $f=dF$, $f^{-1}=dF^*$, $g=dG$, and $g^{-1}=dG^*$ are corresponding subdifferential operators, $A: R^n \to R^m$ is a linear operator; and $A^*: R^m \to R^n$ is the adjoint operator.

This diagram was first considered in [Rockafellar, 1970] which made possible to formulate results in the duality theory of convex programming in maximal generality.

Now let us consider the following four problems.

P1 : Find vectors $(\bar{y}^*, \bar{x}) \in R^m \times R^n$ which make the diagram closed, i.e.
$(\bar{y}^*, \bar{x}): \bar{y}^* \in f(\bar{y}), \bar{x}^* \in g(\bar{x}), \bar{y}=A\bar{x}, \bar{x}^* = A^*\bar{y}^*$.
These relations are called Kuhn-Tucker conditions.

P2 : Find in $R^m \times R^n$ saddle points of the convex-concave Lagrange function ([Rockafellar, 1970, §36])
$La(y^*, x) = G(x) + F^*(y^*) - \langle y^*, Ax \rangle =$
$= G(x) + F^*(y^*) - \langle A^* y^*, x \rangle$

P3,P4 : Solve the dual optimisation problems
$$M(\bar{y}^*) = \sup_x La(y^*, x) = F^*(y^*) - G^*(x^*) \xrightarrow{y^*} \min$$
$$L(\bar{x}) = \inf_{y^*} La(y^*, x) = G(x) - F(y) \xrightarrow{x} \max.$$

The existence of a solution is ensured for problems P1-P4 by Slater conditions in the form $A(ri\, dom\, G) \cap ri\, dom\, F$ is not empty and $A^*(ri\, dom\, F^*) \cap ri\, dom\, G^*$ is not empty.

Main results in the duality theory are summarized by

THEOREM. Under Slater conditions problems P1-P4 have non-empty sets of solutions and are equivalent. Namely,
SOL P1 = SOL P2 = SOL P3 x SOL P4,
where SOL P is the set of solutions of the problem P, and

$$L(\bar{x}) = \max_x L(x) = \min_{y^*} M(y^*) = M(\bar{y}^*).$$

References

Rockafellar, R., <u>Convex Analysis</u>. Princeton, NJ, 1970

Rockafellar, R., Convex Programming and Systems of Elementary Monotonic Relations, <u>J.Math.Anal.Appl.</u>, <u>19</u>, 543-564, 1967

Cundall, P.A., A Computer Model for Simulating Progressive Large Scale Movements in Blocky Rock Systems. In Proc. Symp. of Intern.Soc.of Rock Mech., Nancy,France,1971

Cundall, P.A., Explicit Finite-Dimensional Methods in Geomechanics. In Proc. of the EF Conf. on Numerical Methods in Geomechanics,Blacksburg,Va,1976

Gabrielov A.M.,V.I.Keilis-Borok,T.A.Levshina and V.A.Shaposhnikov, Block Model of Dynamics of the Lithosphere. Math. Methods in Seismology and Geodynamics (Comput.Seismol.,Iss.19),M.:Nauka, 168-177, 1987

Gvishiani A.D. and V.A.Gurvich, Calculation of Equilibrium for Lithospheric Blocks and Engineering Constructions by Convex Programming. Problems of Seismol. Informatics (Comput.Seismol., Iss.21), to appear, 1988

Gvishiani A.D. and V.A.Gurvich, Balanced Flow in Multipoled Scheme. Soviet Phys. Dokl., 29, 1984

Gurvich V.A., The Second Kirchgoff Law for Optimal Traffic Flows. Soviet Phys. Dokl.,29, 1984

Smagin S.A. and A.G.Tumarkin, Boundaries of Reaction of Non-Linear Systems on Seismic Excitations. Problems of Seismol. Informatics (Comput.Seismol., Iss.21), to appear, 1988

GEODETIC MEASUREMENT OF DEFORMATION EAST OF THE SAN ANDREAS FAULT IN CENTRAL CALIFORNIA

Jeanne Sauber

Department of Earth, Atmospheric, and Planetary Sciences, Massachusetts Institute of Technology, Cambridge, MA 02139

Michael Lisowski

U.S. Geological Survey, Menlo Park, CA 94025

Sean C. Solomon

Department of Earth, Atmospheric, and Planetary Sciences, Massachusetts Institute of Technology, Cambridge, MA 02139

Abstract. Triangulation and trilateration data from two geodetic networks located between the San Andreas fault and the Great Valley have been used to calculate shear strain rates in the Diablo Range and to estimate the slip rate along the Calaveras and Paicines faults in central California. The shear strain rates, $\dot{\gamma}_1$ and $\dot{\gamma}_2$, were estimated independently from angle changes using Prescott's method and from the simultaneous reduction for station position and strain parameters using the DYNAP method with corrections to reduce the triangulation and trilateration data to a common reference surface. On the basis of Prescott's method, the average shear strain rate across the Diablo Range for the time period between 1962 and 1982 is 0.15 ± 0.08 μrad/yr, with the orientation of the most compressive strain (β) at N16°E $\pm$ 14°. Utilizing corrections for the deflection of the vertical and the geoid - reference ellipsoid separation computed on the basis of local gravity observations, $\gamma = 0.19 \pm 0.09$ μrad/yr and β = N16°E $\pm$ 13°. Although γ is not significantly greater than zero, at the 95% confidence level the orientation of β is similar to the direction of maximum compressive strain indicated by the orientation of major fold structures in the region (N25°E). We infer that the measured strain is due to compression across the folds of this area; the average shear straining corresponds to a relative shortening rate of 5.7 ± 2.7 mm/yr. In contrast to the situation throughout most of the Coast Ranges where fold axes have orientations approximately parallel to the San Andreas fault, within the Diablo Range between Hollister and Coalinga the trends of the fold axes are different and are thought to be controlled by reactivation of older structures. From trilateration measurements made between 1972 and 1987 on lines that are within 10 km of the San Andreas fault, a slip rate of 10-12 mm/yr was calculated for the Calaveras-Paicines fault south of Hollister. The slip rate on the Paicines fault decreases to 4 mm/yr near Bitter. To distinguish between different models that describe the distribution of strike-slip and compressive displacements within the southern Coast Ranges we compared the findings of regional geologic and geodetic studies with predictions from kinematic plate models. Such comparisons support the view that the fault-parallel component of the San Andreas "discrepancy vector" may be accommodated by strike-slip motion on the Rinconada as well as the San Gregorio fault. Geological and seismicity data, as well as our geodetic results, suggest that northeast-southwest compression in the Coast Ranges of central California may be localized to two regions, the 30-km-wide zone spanned by the triangulation and trilateration network of this study and a second zone to the west of the Rinconada fault. The inferred shortening to the east of the San Andreas fault may represent a significant component of the fault-normal compression predicted by the discrepancy vector.

Introduction

Although most of the relative motion between the Pacific and North American plates in central California is accommodated by slip along the San Andreas fault, distributed compressive and right-lateral strike-slip motion also occurs on faults with surface traces subparallel to the San Andreas fault located between the continental escarpment and the Great Valley [*Gawthrop*, 1977; *Page*, 1981; *Crouch et al.*, 1984; *Eaton*, 1984; *Minster and Jordan*, 1984, 1987; *Namson and Davis*, 1988]. The axis of greatest compression across this region, the southern Coast Ranges subprovince [*Page*, 1981], is thought to be oriented nearly perpendicular to the trend of the San Andreas fault, a result attributed to a combination of slightly convergent relative motion between the Pacific and North American plates and low shear strength along the fault zone [*Mount and Suppe*, 1987; *Zoback et al.*, 1987]. While most recent earthquakes in central California are located on the San Andreas fault (Figure 1), scattered diffuse activity also occurs between the San Andreas fault and the Great Valley. The May 1983 Coalinga earthquake (M_L=6.7), which involved slip on a thrust or reverse fault beneath a young surface fold [*Stein and King*, 1984; *Stein*, 1985], has focused attention on the importance of understanding the state of stress and the rates of deformation to the east of the San Andreas fault in this region and their relation to the overall deformation in the Coast Ranges. In this paper we determine rates of crustal strain in the Diablo Range north of Coalinga from a triangulation and trilateration network and from line-length changes determined by means of trilateration measurements within 10 km of the San Andreas fault. These geodetic results are then compared with other geological and geophysical data to characterize the nature of deformation across the southern Coast Ranges.

Tectonic Setting

The principal strike-slip faults in central California have been well characterized by geologic and geodetic studies. The branched system of subparallel faults near Hollister coalesces southward into a single shear

Copyright 1989 by
International Union of Geodesy and Geophysics
and American Geophysical Union.

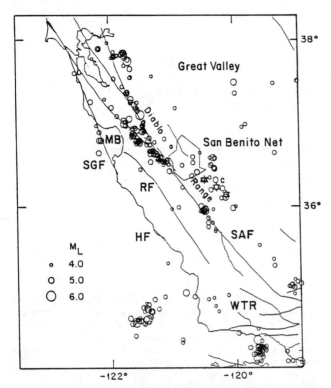

Figure 1. Epicenters of earthquakes ($M_L \geq 4.0$) in the Coast Ranges during 1962 - 1982 [*Engdahl and Rinehart*, 1989]. The locations of three earthquakes with well-determined focal mechanisms, the 1982 Idria, the 1983 Coalinga, and the 1985 North Kettleman Hills events, are indicated by stars. Fault traces are simplified from *Jennings* [1975]: HF = Hosgri fault, RF = Rinconada fault, SAF = San Andreas fault, SGF = San Gregorio fault, WTR =Western Transverse Ranges. An outline of the San Benito triangulation/trilateration network is given for reference. MB = Monterey Bay.

zone, the San Andreas fault, south of Hepsedam (Figure 2). The Calaveras fault is the primary active fault to the northeast of the San Andreas fault near Hollister, while to the southeast of Hollister several faults comprise the southern end of the Calaveras fault zone, among which the Paicines fault is most prominent. Horizontal deformation across the San Andreas and Calaveras-Paicines faults in central California has been measured with near-fault alinement arrays, creepmeters, and trilateration at short and intermediate distances [*Savage and Burford*, 1973; *Thatcher*, 1979a; *Burford and Harsh*, 1980; *Lisowski and Prescott*, 1981]. Between Hollister and Hepsedam (Figure 2), the rate of steady surface slip (creep) across the San Andreas fault increases from ~ 13 to 32 mm/yr and surface slip on the Calaveras-Paicines fault decreases from ~17 to 0 mm/yr. Between Hepsedam and the latitude of Coalinga (Figure 2) near-fault and intermediate-scale geodetic measurements of right-lateral slip are in good agreement and indicate creep at a rate of approximately 32 mm/yr. The rate of slip on this segment of the San Andreas fault estimated from Holocene geological data is 34 ± 3 mm/yr [*Sieh and Jahns*, 1984] at an azimuth of N41°W $\pm$ 2° [*Minster and Jordan*, 1984; *Mount and Suppe*, 1987]. Southward of the latitude of Coalinga shallow slip on the San Andreas fault decreases and the width of the zone of deformation increases over the transition to a locked segment of the San Andreas fault in the Carrizo Plain (the southern aseismic portion of the San Andreas in Figure 1).

The major structural features in the region of this study are shown in Figures 2 and 3. Within 5-10 km to the east of the San Andreas fault the primary geologic structures are related to dextral shear on the San Andreas and Calaveras-Paicines faults (Figure 2). The Diablo Range to the east of this region is a broad antiform which trends approximately N65°W [*Page*, 1985] and encompasses subsidiary fold structures such as the Vallecitos syncline (Figure 3). North of Coalinga the range is pierced by the New Idria diapir (Figure 3) of serpentine and Franciscan rocks. Along the northeast boundary of the study area is the Ortigalita fault (Figure 2), a high-angle fault along or near the contact of Franciscan rocks and the Great Valley sequence to the east [*Raymond*, 1973]. Trenching of the Ortigalita fault zone shows exposed offsets of late Pleistocene and Holocene soils, with possibly as much as 5 km of Quaternary right-slip displacement [*Hart et al.*, 1986].

Multiple phases of deformation in the Diablo Range have been documented by structural analysis [*Namson and Davis*, 1988]. *Harding* [1976] pointed out that there are folds of middle to late Miocene age which are synchronous with the initiation of displacement on the San Andreas fault. The most recent uplift of the Diablo Range began in Pliocene time and most likely accelerated in the Pleistocene [*Page*, 1981; *Page and Engebretson*, 1984]. The Quaternary (< 2.2 Ma) folding is more widely distributed and of much greater structural relief than the Miocene folds [*Namson and Davis*, 1988].

Focal mechanisms of earthquakes in the Diablo Range [*Eaton*, 1985] show a mixture of thrust, reverse, and strike-slip faulting. The locations of three of the larger earthquakes with well-determined focal mechanisms, the October 1982 Idria event (M_L=5.5), the May 1983 Coalinga event (M_L=6.7), and the August 1985 North Kettleman Hills event (M_L=5.5), are given in Figure 1. The focal mechanism determined for the Idria earthquake indicates thrust faulting on a plane oriented N72°E or reverse slip on a plane oriented N64°W [*Eaton*, 1985]. The Coalinga and Kettleman Hills events have similar focal mechanisms with slip occurring on fault planes oriented at about N53°W as either thrusting on a plane dipping shallowly to the southwest or reverse slip on a plane steeply dipping to the northeast [*Eaton*, 1985; J.P. Eaton, personal communication, 1987]. A preliminary focal mechanism determined for an earthquake which occurred on the Ortigalita fault on January 6, 1988 (M_L=3.7), indicates right-lateral slip on a fault plane oriented ~N25°W or left-lateral slip on a plane oriented N65°E (P.A. Reasenberg, personal communication, 1988). The earthquake focal mechanisms and geological structures in the area suggest two primary modes of deformation to the northeast of the San Andreas fault: compression normal to the major fold structures of the region and right-lateral strike-slip motion on faults such as the Calaveras-Paicines and the Ortigalita.

By way of comparison, in the Coast Ranges to the west of the San Andreas fault focal mechanisms [*Gawthrop*, 1977; *Eaton*, 1984; *Dehlinger and Bolt*, 1987] and extensive geological mapping [summarized in *Crouch et al.*, 1984; *Slemmons*, 1987] suggest variable modes of deformation. Between the San Andreas and the Rinconada faults (Figure 1) is the seismically quiescent Salinian block. The upper crust is composed of high-strength granite which is only weakly folded and sparsely faulted [*Dehlinger and Bolt*, 1987; *Slemmons*, 1987]. Focal mechanisms from the Rinconada fault show a mixture of right-lateral strike-slip faulting on northwest striking planes, oblique slip, and reverse faulting [*Gawthrop*, 1977; *Dehlinger and Bolt*, 1987]. From field mapping of recent offsets, D.B. Slemmons (personal communication, 1987) and E.W. Hart (personal communication, 1988) suggest that right-lateral strike-slip faulting dominates the displacement along the Rinconada fault. Between the Rinconada and the San Gregorio-Hosgri faults (Figure 1) the upper crust consists of the very heterogeneous Franciscan complex. Focal mechanisms from this region indicate dominantly oblique reverse faulting along northwest-trending, northeast dipping planes, with P axes oriented N20°-50°E or S20°-50°W [*Dehlinger and Bolt*, 1987]. Significant late-Quaternary right-lateral strike-slip offsets have been measured on the San

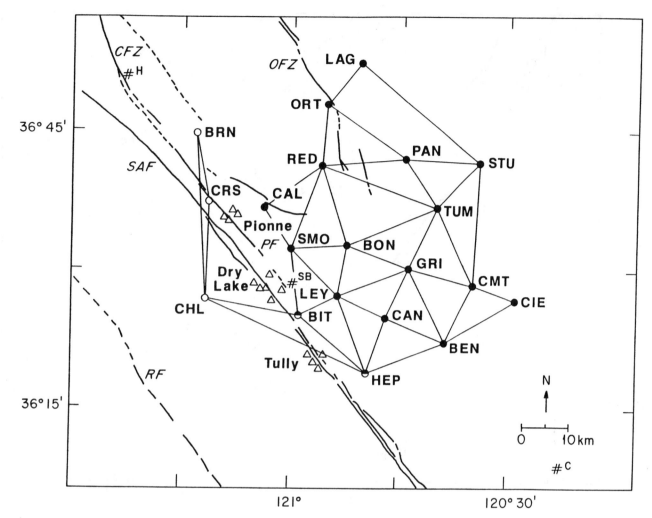

Figure 2. Location of stations in the San Benito triangulation - trilateration network (solid circles), the Coalinga trilateration network (open circles), and small aperture trilateration networks (triangles). Half-filled circles denote stations that were part of both the San Benito and Coalinga networks. Surface traces of Quaternary faults are indicated by solid lines where well located, by dashed lines where approximately located or inferred, and by dotted lines where concealed by younger rocks or by lakes or bays [*Jennings*, 1975]: CFZ= Calaveras fault zone, OFZ= Ortigalita fault zone, PF=Paicines fault, RF=Rinconada fault, SAF=San Andreas fault. Stations discussed in the text include BIT = Bitter, BON = Bonito, BRN = Browns, CHL = Chalone, CRS = Cross, HEP = Hepsedam, LEY = Ley, PAN = Panoche, SMO = Smoker and TUM = Tum. Also shown are the locations of Coalinga (C), Hollister (H), and San Benito (SB).

Gregorio fault [*Clark et al.*, 1984]. South of Monterey Bay this fault branches into several splays, with some branches showing primarily right-lateral strike-slip displacement and others showing east-up reverse faulting [*Hamilton*, 1987]. Focal mechanisms from the San Gregorio-Hosgri fault also indicate right-lateral strike-slip faulting and oblique reverse faulting with a right-lateral component on a northeast dipping plane [*Gawthrop*, 1977; *Eaton*, 1984; *Dehlinger and Bolt*, 1987].

Geodetic Strain Rates

We make use of two geodetic networks to estimate current rates of deformation to the east of the San Andreas fault in central California. The San Benito triangulation and trilateration network spans the Paicines fault zone just east of the San Andreas fault, extends eastward to the western margin of the Great Valley, and is 50 km in extent in the southeast-northwest direction across the Diablo Range (Figure 3). To examine more localized deformation within the zone 10 km to the east of the San Andreas fault, short- and intermediate-range lines from the U.S. Geological Survey (USGS) Coalinga trilateration network were utilized. The geodetic results derived from the Coalinga trilateration data provide an update to the slip rates estimated for the Calaveras and Paicines faults by *Lisowski and Prescott* [1981].

To distinguish between engineering and tensor shear strain, we denote the former by γ and give in units of μrad, whereas the latter is denoted by ε and is given in units of μstrain [*Savage*, 1983]. Uncertainties, where not otherwise stated, are one standard deviation (σ).

San Benito Network

A triangulation survey of the San Benito network was conducted in 1962 by the National Geodetic Survey, and in 1982 a trilateration survey

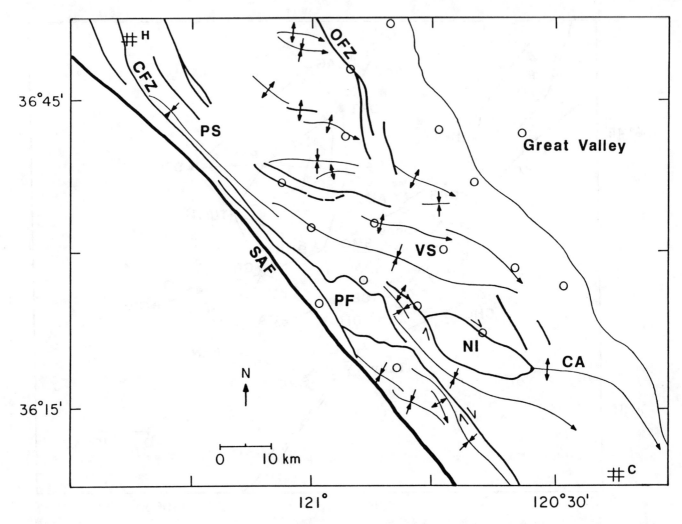

Figure 3. Major structural features of the Diablo Range between Hollister (H) and Coalinga (C), modified from *Dibblee* [1979]. The locations of stations in the San Benito triangulation/trilateration network are shown as circles for reference. Structures shown include: CA = Coalinga anticline, CFZ = Calaveras fault zone, NI = New Idria diapir, OFZ = Ortigalita fault zone, PF = Paicines fault, PS = Paicines syncline, and VS = Vallecitos syncline. Inward pointing double arrows indicate a syncline, outward pointing double arrows an anticline. Single arrows indicate the direction of plunge of a fold axis.

of the San Benito network was performed by the USGS. For interstation visibility, triangulation and trilateration stations are situated on the highest points in a region; thus, many of the stations in the San Benito network are located near anticlinal peaks (Figure 2). To determine the rates of deformation from the San Benito data the observations were processed utilizing two independent procedures: an extended version of Frank's method [*Prescott*, 1976] and the DYNAP method [*Snay*, 1986; *Drew and Snay*, 1988]. In the Appendix we discuss the accuracy of the triangulation and trilateration measurements utilized in this study, the assumptions made in our estimation of the horizontal shear parameters, the considerations involved in combining triangulation and trilateration data to determine the rate of shear strain utilizing the Prescott and DYNAP methods, the corrections applied to reduce the triangulation and trilateration observations to a common reference system, and the methods used to estimate the deflection of the vertical components and the geoid - reference ellipsoid separation needed to make the reduction corrections.

Utilizing both methods we calculate the horizontal shear strain rate components $\dot{\gamma}_1$ and $\dot{\gamma}_2$. In terms of elements ε_{ij} of the strain tensor, $\gamma_1 = \varepsilon_{11} - \varepsilon_{22}$ and $\gamma_2 = \varepsilon_{12} + \varepsilon_{21}$, where the strain tensor is referred to a geographic coordinate system in which the 1-axis is directed east and the the 2-axis is directed north. The strain component γ_2 is equal to the decrease induced by strain in the right angle between northward- and eastward-directed lines, whereas γ_1 is equal to the increase in the angle between lines directed northwest and northeast. Results are generally given in terms of the maximum shear strain rate $\dot{\gamma}$, where $\dot{\gamma}^2 = \dot{\gamma}_1^2 + \dot{\gamma}_2^2$, and the orientation ψ of the vertical plane with maximum rate of right-lateral shear [*Frank*, 1966; *Prescott*, 1976]. For comparison with the trends of fold structures of the Diablo Range, the orientation β of the maximum rate of compressive strain is sometimes given instead of the orientation of maximum rate of right-lateral shear.

Results with Prescott's method. The observations used to estimate $\dot{\gamma}_1$ and $\dot{\gamma}_2$ in the extended version of Frank's method [*Prescott*, 1976] are

TABLE 1. Strain Rate Parameters for Spatial Subsets of the San Benito Network

Subnet	Number of Angles	$\dot{\gamma}_1$, μrad/yr	$\dot{\gamma}_2$, μrad/yr	$\dot{\gamma}$, μrad/yr	β	ψ
East	14	0.18 ± 0.11	0.01 ± 0.12	0.18 ± 0.10	N 1°W ± 20°	N46°W ± 20°
West	9	0.07 ± 0.14	-0.18 ± 0.13	0.19 ± 0.13	N34°E ± 22°	N11°W ± 22°
North	13	0.04 ± 0.10	-0.12 ± 0.10	0.16 ± 0.11	N37°E ± 20°	N 8°W ± 20°
South	7	0.22 ± 0.16	-0.13 ± 0.22	0.25 ± 0.20	N10°E ± 20°	N35°W ± 20°
All stations except BIT, HEP, and PAN	25	0.13 ± 0.08	-0.08 ± 0.08	0.15 ± 0.08*	N16°E ± 14°*	N29°W ± 14°*

* These strain parameters have been scaled by the a posteriori variance factor [*Vanicek and Krakiwsky*, 1986].

changes in angles. In this study we determined angle changes from two different data types, reflecting the different types of surveys made in 1962 and 1982. Further, we compared angles measured on the Earth's surface to angles determined from a network adjustment on a reference ellipsoid. In using data derived from different measurement techniques it is preferable to reduce the data to a common reference surface. The required reduction corrections are discussed in the Appendix. In employing Prescott's procedure to determine the strain parameters we did not make these corrections. In our use of the alternative DYNAP method these corrections are performed, and we compare the results from the two techniques to illustrate in part the utility of making these corrections.

The angle changes associated with the stations Bitter, Hepsedam, and Panoche were significantly larger than angle changes from other portions of the network and were, therefore, examined separately. Bitter and Hepsedam are located near the Paicines fault zone. Using only the 11 angle measurements that include one of the stations Bitter or Hepsedam gives $\dot{\gamma} = 0.56 \pm 0.16$ μrad/yr and $\Psi = $ N28°W ± 11°. The shear strain rate across the approximately 10-km-wide zone to the northeast of stations Bitter and Hepsedam, assuming right-lateral motion on the Paicines fault, implies a rate of slip of 5 ± 2 mm/yr. Additional geodetic data from the Coalinga trilateration network relating to deformation across the San Andreas and Calaveras-Paicines faults are discussed in the next section.

The relatively large angle changes around station Panoche are not so easily explained. Using only the seven angle changes that include Panoche gives $\dot{\gamma} = 0.76 \pm 0.27$ μrad/yr and $\psi = $ N58°W ± 9°, or $\beta = $ N13°W ± 9°. These results are not consistent with either shear strain across the Ortigalita fault zone or contraction across the major fold structures of the region.

The strain rate parameters estimated on the basis of 25 angle changes in the central portion of the San Benito network, excluding angles to Bitter, Hepsedam, or Panoche, were $\dot{\gamma} = 0.15 \pm 0.08$ μrad/yr and $\beta = $ N16°E ± 14°, or $\psi = $ N29°W ± 14°. The standard deviations reflect both misfit and data uncertainties due to measurement imprecision.

To search for significant strain inhomogeneity, the shear strain parameters were estimated from spatial subsets of these data, again excluding the Bitter, Hepsedam, and Panoche stations (Table 1). There is a trade-off between improving the precision of the strain estimate by using a larger number of angles and averaging spatial variations as the size of the sampled region is increased. The data were first broken into two distinct groups, one set closer to the Great Valley ("east") and one set closer to the San Andreas fault ("west"). If the measured rates of shear strain were due to strain accumulation on the adjacent San Andreas fault, the rate of shear strain would be higher in the western subnet. Alternatively, if the rate of compressive strain was higher across the folds near the Great Valley, the eastern subnet might show a higher shear strain rate. There is no suggestion of a significant rate difference, however, between the strain rates determined from the eastern and western data subgroups. The set of 25 angles were also divided into "north" and "south" sets to look for any change which might be associated with along-strike variations on the San Andreas and Calaveras-Paicines faults. Although the strain rate results differ in the subnets, particularly in the orientation of maximum rate of compression, the magnitudes and orientations of strain rate in the various subregions are not significantly different from the average values determined from the complete set of 25 angles.

Results with the DYNAP method. In a second approach, the directions observed in 1962 and the distances measured in 1982 were employed to solve simultaneously for crustal motion parameters and the horizontal positional coordinates of the geodetic stations using the DYNAP method [*Snay*, 1986; *Drew and Snay*, 1988]. With DYNAP we are able to make single epoch adjustments to search for any observational errors during the individual surveys, and we are able to evaluate the effect of making reduction corrections to the distance and direction observations.

We first performed separate network adjustments for 1962 and 1982. A particular solution was determined by holding the minimum number of free parameters fixed. For the two-dimensional adjustment of 1962 triangulation data, these free parameters are two components of translation, a network rotation, and scale. For the two-dimensional adjustment of the 1982 trilateration data, scale is not a free parameter. Examination of the largest standardized residuals (given by the difference between observed and calculated values divided by the standard deviation of the observation) indicates that the 1962 direction observations to the station Panoche have large residuals. This may account for the anomalously large angle changes associated with Panoche discussed earlier. Large residuals are not systematically associated with any one station in the 1982 adjustment.

From the analysis utilizing the Prescott method we obtained the result that the angle changes associated with the stations Bitter and Hepsedam were consistent with right-lateral slip at the approximate orientation of the Paicines fault. To test this result we solved for strain parameters and horizontal coordinates for a small subnetwork consisting of Bitter, Hepsedam, Smoker, and Ley (Figure 2). The rate of shear strain was estimated to be 0.53 ± 0.35 μrad/yr with $\psi = N40°W \pm 19°$, similar to the orientation of the San Andreas and Paicines faults.

The shear strain rate parameters $\dot{\gamma}_1$ and $\dot{\gamma}_2$ were estimated by the DYNAP method using data from the central portion of the network (excluding measurements to Panoche, Bitter and Hepsedam). Without taking into account corrections the rate of shear strain $\dot{\gamma} = 0.15 \pm 0.09$ μrad/yr and $\beta = N17°E \pm 16°$, similar to the result obtained utilizing Prescott's method. The rate of shear strain determined using the corrected data, denoted by $\dot{\gamma}_c$ is 0.19 ± 0.09 μrad/yr, with $\beta = N16°E \pm 13°$ (Figure 4).

As shown in Figure 4, at the 95% confidence level $\dot{\gamma}$ is not significantly greater than zero and β is not significantly different from the orientation predicted for shear strain associated with slip on the San Andreas fault (N4°E). The orientation of β, however, is similar to the direction of maximum compressive strain indicated by the orientation of major fold structures in this region. If we assume that the strain represents uniform horizontal contraction across a 30-km-wide region in the direction N16°E and that there is no extension in the orthogonal direction, this average strain rate corresponds to 5.7 ± 2.7 mm/yr of shortening.

The error in the estimate of shear strain rate determined in this study is dominated by the less accurate triangulation survey (see Appendix). It is instructive to estimate the accuracy that could be achieved with an additional trilateration survey of this network. An additional survey of all the stations in the San Benito network in 1992, for instance, would provide ~0.02 μstrain/yr accuracy in the principal strains, and all of the horizontal components of the tensor rate of strain ($\dot{\varepsilon}_{ij}$) could be estimated. At this level of accuracy we could better constrain the rate of crustal deformation across the Diablo Range as well as discern spatial variations in the deformation field.

Coalinga Trilateration Network

We have employed short and intermediate-length lines from the Coalinga trilateration network (Figure 2) to update previous estimates [*Lisowski and Prescott*, 1981] of the slip rates on the Calaveras and Paicines faults. Since 1972 trilateration measurements have been made periodically by the USGS in central California on a regional scale spanning a 20-km-wide zone centered on the San Andreas fault and including several distinct faults, as well as on smaller (1-2 km) aperture networks that span a single fault (Pionne, Dry Lake and Tully nets). The estimated accuracy in the line lengths determined from the short-range measurements is 4 mm [*Lisowski and Prescott*, 1981]. The estimates of line-length change made by *Lisowski and Prescott* [1981] also included earlier data collected by the California Division of Mines and Geology, but we have utilized only the more homogeneous set of USGS data.

For the region within 10 km to the east of the San Andreas fault we assume that the predominant mode of crustal deformation is right-lateral fault slip on the Calaveras-Paicines fault system. This is equivalent to the assumption that the measured deformation in this region is due neither to elastic strain accumulation that will be released episodically in earthquakes on the adjacent San Andreas fault nor to crustal shortening normal to the San Andreas fault. Several lines of reasoning support this assumption. As discussed above, on the segment of the San Andreas fault adjacent to the San Benito network, fault slip occurs primarily by steady creep, and we expect little right-lateral shear strain accumulation associated with the San Andreas fault to be measured on off-fault geodetic lines. In addition, elastic strain accumulation associated with the San Andreas fault would be measurable geodetically on both sides of the fault, yet to the west of the San Andreas fault the maximum right-lateral shear

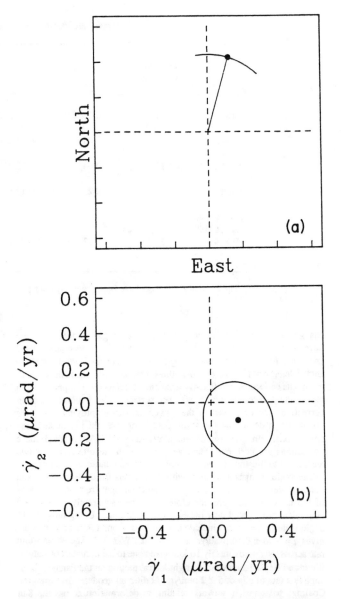

Figure 4. Strain rate inferred for the San Benito network by use of the DYNAP method including corrections for deflection of the vertical and separation of the geoid and reference spheroid. (a) The orientation (β) of the axis of maximum compressive strain. The uncertainty in azimuth, at 95% confidence, is shown by an arc. (b) The shear strain rate parameters $\dot{\gamma}_1$ and $\dot{\gamma}_2$ together with a 95% confidence ellipse.

strain within the Salinian block, estimated from triangulation data measured during the time interval 1944-1963, is poorly resolved and is not oriented parallel to the San Andreas fault [*Thatcher*, 1979a]. An alternative mode of deformation adjacent to the San Andreas is compression across structures such as the Paicines syncline (Figure 3). Strike-slip motion across this region would shorten north-south lines and extend east-west lines, whereas contraction would shorten northeast-southwest lines. Line length changes in the Pionne short-range network and the lines from Bitter to Hepsedam and Browns to Cross are consistent

TABLE 2. Fault-Slip Rates Inferred from Length Changes on Fault-Crossing Lines

Calaveras - Paicines Fault Zone

Station 1	Station 2	No. of Obs.	Period	$\dot{L}$, mm/yr	Azimuth, deg	Slip Rate, mm/yr
BRN	CRS	7	72.7-83.9	-8.7 ± 0.8	168	9.9 ± 0.9
CRS	CHL	9	72.1-83.9	-18.5 ± 0.4	181	24.9 ± 0.5
BRN	CHL	8	72.1-83.9	27.4 ± 0.4	175	33.8 ± 0.5

Paicines Fault Adjacent to the Central Creeping Portion of the San Andreas Fault

Station 1	Station 2	No. of Obs.	Period	$\dot{L}$, mm/yr	Azimuth, deg	Slip Rate, mm/yr
BIT	HEP	6	73.1-83.9	3.7 ± 0.5	129	3.8 ± 0.5
CHL	BIT	8	72.1-83.9	19.8 ± 0.7	280	25.8 ± 0.9
CHL	HEP	7	72.7-83.9	27.4 ± 0.6	113	30.7 ± 0.6

The quantity $\dot{L}$ is the rate of change of line length determined by least-squares, shown together with one standard deviation. Azimuth is measured clockwise from north. The slip rate is that appropriate to the strike-slip fault crossed by the indicated line. The time intervals of observations are given in decimal fractions of years.

with the hypothesis of right-lateral slip on the Calaveras-Paicines fault system. As discussed above, results from the San Benito geodetic subnetwork that includes measurements to Bitter and Hepsedam are most consistent with right-lateral shear at the orientation of the San Andreas and Paicines fault.

If the deformation in this region occurs as rigid block motion along the Calaveras-Paicines fault system, the observed length changes can be converted into a fault-parallel displacement rate from the slope of the least-squares linear fit to the interstation length data [*Prescott and Lisowski*, 1983]. The slip rates determined from lines crossing the fault at a low angle are given in Table 2 for the Calaveras-Paicines region just south of Hollister and for the Paicines fault zone between Bitter and Coalinga. The sum of the slip rate determined from the station pairs that cross only the San Andreas fault (Cross-Chalone and Chalone-Bitter lines) and the slip rate from pairs that are completely east of the San Andreas fault (Browns-Cross and Bitter-Hepsedam) are approximately equal to the slip rate from station pairs that span both fault zones (Browns-Chalone and Chalone-Hepsedam lines). The line length changes across the Calaveras-Paicines fault zone are consistent with ~10 mm/yr of right-lateral slip. The rate of slip estimated for the Paicines fault south of the city of San Benito is 4 ± 1 mm/yr.

The Bitter to Hepsedam line showed a significant change in slope beginning in mid-1978. This change corresponded to an increase in the rate of slip on the Calaveras-Paicines system and a decrease in slip rate on the San Andreas fault. A similar change after 1979 in the slip rates for the Calaveras and San Andreas faults has been inferred from geodetic measurements of the USGS Hollister trilateration network [G. Gu and J. C. Savage, personal communication, 1986; *Matsu'ura et al.*, 1986]. The increase in slip on the Calaveras-Paicines fault coincided approximately with the occurrence of the 1979 Coyote Lake earthquake (M_L=5.7) on the northern Calaveras fault. The rate of line length change, $\dot{L}$, between Bitter and Hepsedam for the 1973-1978 time period was 1 ± 1 mm/yr [*Lisowski and Prescott*, 1981]. The higher rate of slip given in Table 2 thus is largely due to the increased rate during 1978-1984.

The average fault slip indicated by the three short-range networks (Figure 2), again assuming simple block motion, is given in Table 3. The rate of slip on the Pionne net, 12 ± 2 mm/yr, is similar to the rate estimated from the line between Cross and Browns which crosses the

TABLE 3. Fault Slip Rates Indicated by Short-Range Trilateration Networks

Network	No. of Observations	Period	Slip Rate, mm/yr
Pionne	8	75.2-87.3	12 ± 2
Dry Lake	4	79.0-87.3	27 ± 2
Tully	8	74.9-87.3	32 ± 1

See Table 2 for explanation of notation.

Calaveras-Paicines fault zone (10 ± 1 mm/yr). Between the Dry Lake and Tully networks the rate of creep on the San Andreas fault increases from 27 ± 2 to 32 ± 1 mm/yr. South of the Tully network the rate of slip determined from near-fault data on the San Andreas fault is the same as the rate estimated from the Chalone to Hepsedam line [*Prescott and Lisowski*, 1981].

The short and intermediate-range trilateration measurements document a transfer of slip associated with the Calaveras-Paicines fault zone to the San Andreas fault. The narrowing of the region that accommodates the ~32 mm/yr slip rate also corresponds to a transition from a complex multi-stranded portion of the fault system with locked segments that break in periodic moderate-to-large earthquakes to a geometrically simpler segment that accommodates slip through creep.

Comparison with Other Geodetic and Geologic Observations

As noted above, the strain rate parameters estimated on the basis of angle changes in the central portion of the San Benito network are γ_c = 0.19 ± 0.09 μrad/yr and β = N16°E ± 13°, or ψ = N29°W ± 13°. Interpreted in terms of uniform horizontal contraction in the direction given by the angle β, the rate of shortening is 5.7 ± 2.7 mm/yr. Although there is no significant strain at the 95% confidence limit, the orientation of the principal strain directions are consistent with the geological structures of the region (Figure 3). The azimuth of the least compressive strain (N74°W ± 13°) is close to the trend of the major fold structures of the region (N65°W). The direction of maximum rate of right-lateral shear is also close to the trend of the major strike-slip faults of the region.

From a geological reconstruction of the structures between the Great Valley and the San Andreas fault at the approximate latitude of Coalinga, *Namson and Davis* [1987] inferred that 11 km of late Cenozoic shortening has occurred perpendicular to the San Andreas fault. If active folding commenced 5 m.y. ago, the average rate of shortening has been 2.2 mm/yr. If folding began as recently as 2 m.y. ago an average rate of shortening of 5.5 mm/yr is implied. These figures are comparable to the geodetically inferred rate.

The rate of slip obtained in this study for the Calaveras-Paicines fault zone south of Hollister is 10 ± 1 mm/yr (Browns to Cross, Table 2) and 12 ± 2 mm/yr (Pionne net, Table 3). These slip rates may be compared to other estimates of slip rate derived from geological and geodetic observations. *Harms et al.* [1987] inferred a late Quaternary slip rate on the Paicines fault of 3.5-13 mm/yr from an offset terrace riser and an offset hill ~10 km south of Hollister. Between 1973 and 1986 the average rate of slip obtained from offset of the USGS Thomas Road alinement array, which spans the Paicines fault near its intersection with the Browns-Cross line, is approximately 6 mm/yr [*Harsh and Pavoni*, 1978; S. Burford, personal communication, 1988]. From trilateration measurements made between the stations Browns and Cross during an interval earlier than but overlapping that of our study (69.8-78.4), *Prescott and Lisowski* [1981] determined a slip rate of 8 ± 1 mm/yr. *Prescott and Lisowski* [1981] also calculated the rate of the slip from short-range trilateration measurements of the Pionne net to be 10 ± 3 mm/yr for approximately the same period. These estimates are not significantly different from the results reported here. Utilizing data from the (USGS) Hollister trilateration network, located north of our study region, *Matsu'ura et al.* [1986] inverted for fault displacement rate versus depth on the San Andreas, Calaveras, and Sargent faults. They estimated the rate of slip on the Calaveras-Paicines fault between 1971 and 1983 and between latitudes 36.70° and 36.87°N to be 18 ± 4 mm/yr, with no significant surface creep. This result is higher than the rate of surface slip documented from the Thomas Road alinement array as well as with that obtained in this study. The inconsistency of the *Matsu'ura et al.* [1986] rate with other results from the region may be due to the fact that the Hollister network does not span the southern portion of the Calaveras fault; in particular, the 18 mm/yr value may reflect an average rate over a region where slip on the Calaveras fault is decreasing to the southeast.

The rate of slip obtained in this study for the Paicines fault south of the city of San Benito is 4 ± 1 mm/yr (Bitter to Hepsedam, Table 2). While geological mapping indicates that recent right-lateral slip has occurred on the Paicines fault as far south as the city of San Benito, no active fault trace has been mapped further south (J. Perkins, personal communication, 1988). Sedimentation rates are high in this region, however, so a low rate of slip on the fault may be masked by alluvium. From a study of earthquake focal mechanisms and seismicity of this region, *Ellsworth* [1975] suggested that slip on the San Andreas fault is transferred to the faults northeast of the San Andreas between Bitter and Cross (Figure 2). The rate of slip estimated from the change in the Bitter-to-Hepsedam line length, 3-4 mm/yr, however, suggests that some slip may be occurring on an extension of the Paicines fault south of San Benito. As discussed above, the slip rate inferred for this line is higher than the rate of slip of 1 ± 1 mm/yr estimated by *Lisowski and Prescott* [1981] for an earlier period and may indicate a temporal variation in the rate of deformation in this region.

Principal Directions of Strain and Stress

The principal directions determined geodetically for rate of strain within the San Benito network may be compared with the orientation of the maximum principal stress (σ_1) estimated from wellbore breakouts and the azimuths of P axes determined from earthquake focal mechanisms for the region east of the San Andreas fault. As noted earlier, the orientations of principal stress directions determined from breakout orientations, earthquake fault plane solutions, and the azimuths of major fold structures in central California have been used to distinguish between models relating the formation of fold structures in the southern Coast Ranges to motion along the San Andreas fault and to infer the state of stress on and near the San Andreas fault [*Zoback et al.*, 1987; *Mount and Suppe*, 1987; *Namson and Davis*, 1988].

Due to extensive drilling for oil to the east of the San Andreas fault, a large number of wells have been available for measurement of stress-induced wellbore breakout orientations. Borehole breakouts are caused by unequal stress concentrations around a borehole wall and create elongations of the hole in directions perpendicular to the orientation of the maximum horizontal stress [*Springer*, 1987]. Breakouts are generally oriented northwest-southeast, indicating a northeast-southwest orientation for the greatest compressive stress, a direction perpendicular to the axes of major folds near the Great Valley and at a 70°-90° angle to the San Andreas fault [*Springer*, 1987; *Zoback et al.*, 1987; *Mount and Suppe*, 1987]. To the east of the San Andreas fault the majority of breakout measurements have been made in wells near the western edge of the Great Valley (Figure 5); comparatively few wellbore breakout orientations have been obtained closer to the San Andreas fault within the area of the geodetic measurements reported in this paper.

The orientation of maximum principal stress may also be inferred from well-determined fault plane mechanisms, notably from the 1982 Idria, the 1983 Coalinga, and the 1985 North Kettleman Hills earthquakes (Figure 5). For the Coalinga and the North Kettleman Hills events the P-axis orientation is ~N37°E [*Eaton*, 1985; J.P. Eaton, personal communication, 1987], similar to the directions of σ_1 inferred from breakout orientations in the same region. The azimuth of the P axis for the Idria event is N12°E [*Eaton*, 1985].

As may be seen in Figure 5, the breakout orientations and the P-axis directions are very similar to the orientation of maximum compressive strain implied by the trend of local fold structures and by the inferred direction of horizontal shortening across the San Benito network. Along the western edge of the Great Valley and near Coalinga, the direction of σ_1 inferred from the breakout orientations and two earthquake focal mechanisms and the orientation of maximum compressive strain inferred

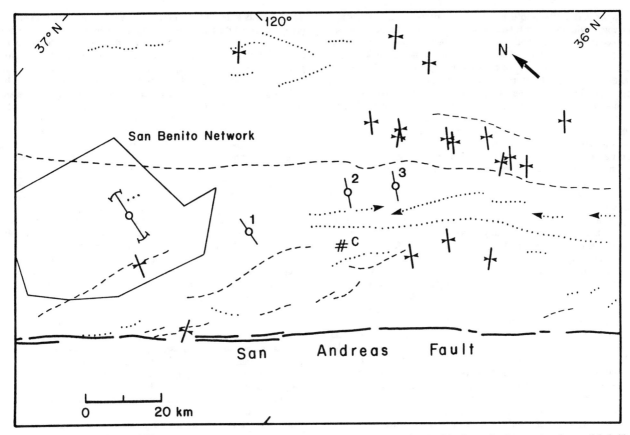

Figure 5. Fold structures and measured stress orientations in the region east of the San Andreas fault in central California [modified from *Mount and Suppe*, 1987]. Synclines are indicated by dashed lines, anticlines by dotted lines. Single arrows indicate the direction of plunge of a fold axis. Wellbore breakout measurements are given by a solid line with inward pointing arrows perpendicular to the direction of σ_1. The direction of σ_1 inferred from the azimuth of the P axes for three earthquakes are labeled by number: 1 = 1982 New Idria earthquake, 2 = 1983 Coalinga earthquake, and 3 = 1985 North Kettleman Hills earthquake. The axis of maximum compressive strain (β) determined from triangulation and trilateration data from the San Benito network is given along with its standard deviation.

from the fold trends is N30-50°E. To the northwest of Coalinga, including the region spanned by the San Benito network, the direction of maximum compressive strain predicted from the azimuth of the local fold structures, the σ_1 direction inferred from one wellbore breakout orientation, and the P-axis orientation of the Idria earthquake are N12-35°E. While *Mount and Suppe* [1987] have suggested that the folds in this area have been inactive since the Miocene, the rate of deformation determined for the region spanned by the San Benito network suggests that ongoing compressive strain is being accommodated across these folds. The recency of folding in this region is further supported by the geological observation that the folds deform Quaternary deposits [*Namson and Davis*, 1988].

As summarized by *Mount and Suppe* [1987] and *Namson and Davis* [1988] the occurrence of folding in the Coast Ranges has been variously attributed to distributed right-lateral shear associated with the San Andreas fault or to oblique displacement across the region. On the basis of experimental simulations of simple shear and field studies of wrench faulting, en echelon fold structures are predicted to occur adjacent to a strike-slip fault due to distributed shear [*Wilcox et al.*, 1973]. Fold axes are expected to be oriented perpendicular to σ_1 in the early stages of wrench faulting; the folds may subsequently rotate into parallelism with the strike-slip fault through distributed shear [*Mount and Suppe*, 1987].

In central California, the axes of early-forming en echelon folds would be at an angle of 30° ± 15° to the trend of the San Andreas fault (~N41°W) in a counterclockwise direction, or N71°W ± 15° (β = N19°E ± 15°).

In an alternative model deformation is due to oblique displacement, sometimes termed "transpression"[*Harland*, 1971], across the southern Coast Ranges. This deformation is thought to be decoupled into a low-shear-stress, strike-slip component and a high-deviatoric-stress, thrust component [*Mount and Suppe*, 1987; *Zoback et al.*, 1987]. In this model the strike-slip component is accommodated within a narrow (< 3-10 km wide) zone and the compressive component is accommodated over a wider zone (10-100 km). The compression is held to be at least partly the result of a small (~ 6°) difference between the orientation of the San Andreas fault in central California and the local direction of Pacific-North American relative plate motion [*DeMets et al.*, 1987]. The precise orientation of σ_1 adjacent to the San Andreas fault, according to this model, depends on the relative strength of the fault zone and the surrounding lithosphere [*Mount and Suppe*, 1987; *Zoback et al.*, 1987]; for a substantially weaker fault zone σ_1 may be nearly normal to the fault zone (β ~ N49°E).

The direction of maximum compressive strain indicated by orientations of the local fold structures in the San Benito region is N25°E, close to the orientation of maximum incremental compressive strain estimated from the geodetic data (β = N16°E ± 13°). These orientations are in apparent agreement with the wrench faulting model. The significant difference between the trend of the fold axes in the San Benito region and in the San Emigdio Mountains [*Davis*, 1986] and the strike of the San Andreas fault is in contrast, however, to the situation throughout most of the Coast Ranges in central California where fold axes have orientations approximately parallel to the San Andreas [*Mount and Suppe*, 1987]. In these two regions the fold orientations are thought to be controlled by reactivation of older structures (T.L. Davis, personal communication, 1988); in the San Benito region these older structures may be related to a late Miocene deformation event [*Namson and Davis*, 1988]. Given such structural control, the geodetic data reported here are also consistent with the oblique displacement, or transpression, model for deformation of the Coast Ranges.

As discussed earlier there are two observations which argue against models in which right-lateral shear strain is distributed across a zone significantly greater than 10 km in width. Because slip on the adjacent San Andreas fault occurs primarily by steady creep, little of the right-lateral shear strain accumulation associated with fault should be measurable on off-fault geodetic lines. Additionally, distributed shear strain associated with the San Andreas fault would be observable on both sides of the fault, yet there is no geodetic or geologic evidence of deformation within the Salinian block [*Thatcher*, 1979a].

There are several factors which may complicate the interpretation of measurements used to distinguish between models relating the formation of fold structures in the Coast Ranges to motion along the San Andreas fault. First, there may be processes that act on different length scales which might explain the change in the σ_1 direction and the trend of fold structures at different distances from the San Andreas fault. Within a narrow zone centered on the San Andreas fault, local geological structures and focal mechanisms may be the result of geometrical complexities in the fault trace [*Segall and Pollard*, 1980] or the rheological structure of the fault zone [*Horns et al.*, 1985]. Between the San Andreas fault and the Paicines fault, for example, the focal mechanisms determined by *Ellsworth* [1975] may be due to interaction between the two faults. As seen in the San Benito region, fold orientations may also vary due to reactivation of older structures or material heterogeneity. Finally, as will be discussed in the next section, there is the additional complexity that deformation in the southern Coast Ranges is related, in a kinematic sense, to the overall deformation across the Pacific-North American plate boundary.

Relation of Deformation East of the San Andreas Fault to the Accommodation of Plate Motion

The results of the geodetic measurements presented in this paper are relevant to deformation in other subregions of the Coast Ranges and to the question of how Pacific-North American plate motion is accommodated across California.

Global plate motion models, which yield the relative motion between the North American and Pacific plates, have been used as kinematic boundary conditions on the integrated deformation across the plate boundary zone in the western United States [*Minster and Jordan*, 1984, 1987; *Bird and Rosenstock*, 1984; *Weldon and Humphreys*, 1986]. If the San Andreas fault functioned as a simple boundary that accommodated the full motion between two rigid plates, the rate of slip in central California predicted by the NUVEL-1 global plate motion model would be about 49 mm/yr at N35°W [*DeMets et al.*, 1987]. Deformation across the Pacific-North American boundary is instead thought to be distributed across a broad zone between the continental escarpment and the eastern front of the Basin and Range province. As summarized above, the rate of slip on the San Andreas fault in central California oriented at N41°W ± 2° is constrained from Holocene geological data and ground-based geodetic measurements to be 34 ± 3 mm/yr. The vector difference between the plate motion and the San Andreas rate is about 16 mm/yr in the direction N21°W. The integrated rate of extensional deformation to the east of the San Andreas fault across the Basin and Range has been estimated from geological observations averaged over the Holocene and from very long baseline interferometry (VLBI) to be 9.7 ± 2.1 mm/yr at N56°W ± 10° [*Minster and Jordan*, 1987]. The vector difference derived using the above rates of motion for the San Andreas fault and Basin and Range, referred to as the discrepancy vector by *Minster and Jordan* [1987], is about 10 mm/yr in the direction N14°E, or 5 mm/yr of slip parallel to the San Andreas fault and 8 mm/yr of convergence normal to the fault. On the basis of the estimates of *Minster and Jordan* [1987], the uncertainties in the discrepancy vector are approximately ± 5 mm/yr for the rate of slip and ± 15° for the direction. Although some minor internal deformation within the Sierra Nevada block [*Lockwood and Moore*, 1979] or across the Great Valley syncline may occur, most of the deformation represented by the discrepancy vector is thought to occur within the Coast Ranges. *Weldon and Humphreys* [1987] and *Saucier and Humphreys* [1988] have constructed a self-consistent description of deformation in southern and central California from geodetic data and Quaternary geologic slip rates. From their model they predict that the Pacific-North American relative plate motion vector is 9° more westerly than that given by NUVEL-1. They estimate ~ 5 mm/yr of convergence normal to the San Andreas fault in central California.

A compilation of estimated and measured rates of deformation in the Coast Ranges, separated into right-lateral strike-slip motion on specific faults and distributed compression, is given in Table 4. In general the geological data are most useful for indicating long-term modes of deformation and for placing upper and lower bounds on rates. The large range in most geologically determined rates is due to uncertainties in dating rock units and in the timing of geological reconstructions.

A comparison of incremental strain rates determined over a geologically short interval to long-term slip rates is meaningful only if the deformation process accumulates strain in a temporally uniform manner. On short-time scales (1-100 yrs) there are a number of mechanisms associated with temporal variations in strain rate. An increase in the rate of shear strain has been documented to occur after large earthquakes in a region adjacent to the rupture zone [*Thatcher*, 1986]. There is no evidence, however, for a change in the rate of slip on the San Andreas fault in central California for the time interval that spans the occurrence of the great 1906 earthquake on the northern locked segment of the San Andreas fault [*Thatcher*, 1979a]. As discussed above, there is a suggestion that the rate of slip on the Paicines and Calaveras faults changed at about the time of the 1979 Coyote Lake earthquake.

Within a fold-and-thrust belt such as the southern Coast Ranges the short-term temporal variations in the rate of deformation are poorly constrained. The pattern of moderate size earthquakes (M_L > 5.0) between 1932-1982 in the southern Coast Ranges [*Engdahl and Rinehart*, 1989] is similar to the seismicity pattern given in Figure 1; there is some additional seismicity in the region between the Rinconada fault and the continental escarpment (Figure 1). This result suggests that over the past ~50 years the pattern of release of stress and strain by earthquakes has been approximately time-stationary. Comparison with geological data is further complicated by the observation that the plate boundary zone undergoes evolution on geological times scales. Field geological evidence from the Coalinga region indicates that the most recent episode of uplift began only 2-3 m.y. ago, well after slip commenced on the San Andreas fault [*Namson and Davis*, 1988]. A fold-and-thrust belt typically undergoes an evolution such that the locus of deformation changes. In the initial stages of deformation the locus of activity is controlled by such factors as the nature and position of the driving and resisting forces, pre-existing zones of weakness, and rock properties such as the elastic moduli, effective viscosity, and stratification of the deforming medium

$$x(t_i) = x(t_0) + (t_i - t_0) L x(t_0) \quad (A.2)$$

where $x(t_i)$ are the two dimensional station coordinates at time t_i and t_0 is the reference time. The four components of the 2 x 2 tensor L parameterize the velocity field v by its gradient [*Malvern*, 1969]:

$$L_{ij} = \frac{\delta v_i}{\delta x_j} \quad (A.3)$$

In equation (A.2) the origin is arbitrary. In practice one station is chosen as an origin and is held fixed for all epochs. L can be decomposed into a sum of a symmetric tensor D, called the rate of deformation tensor, and a skew-symmetric tensor W, called the spin tensor [*Malvern*, 1969]

$$L = D + W \quad (A.4)$$

For the case where all of the components of rate of deformation are zero, the instantaneous motion is then a rigid-body rotation. When displacements and displacement gradients are small D is approximately equal to $\dot{\varepsilon}_{ij}$.

Corrections to Reduce the Observations to a Common Reference System

When a direction measurement is made the theodolite is leveled; thus the measurement is made normal to the geoid, not normal to a reference surface. A correction for the deflection of the vertical is therefore required [*Bomford*, 1980; *Vanicek and Krakiwsky*, 1986]. The deflection of the vertical is the spatial angle between the vector normal to the geoid and the vector normal to an ellipsoidal surface. The correction $\Delta\alpha_{ij}$ to a direction of azimuth α_{ij} and elevation angle v_{ij} [*Bomford*, 1980] is

$$\Delta\alpha_{ij} = -[\xi_i \sin \alpha_{ij} - \eta_i \cos \alpha_{ij}] \tan v_{ij} \quad (A.5)$$

where ξ_i and η_i are the meridian and prime vertical components of the deflection of the vertical at the observing, or ith, station. Deflections of the vertical are usually estimated from astronomical azimuth observations or from computed values based on local gravity observations [*Coleman and Lambeck*, 1983]. Alternatively, the deflections can be computed from the long-wavelength part of the geopotential together with necessary transformation parameters between the geoid and the ellipsoid reference system.

In this study the initial latitude and longitude of the station are positions given in the NAD83 reference system [*Defense Mapping Agency*, 1987]. The initial heights are orthometric heights, the height above the geoid estimated primarily from spirit leveling. For the station coordinates to be in a common reference frame, the heights need to be converted to heights above the GRS 80 reference ellipsoid [*Defense Mapping Agency*, 1987]. To make this calculation the separation of the geoid and the reference ellipsoid needs to be estimated. Since astronomical azimuths are not available, the deflection of the vertical and the geoid - ellipsoid separation are initially estimated from the long-wavelength part of the geopotential.

The global gravity field representation given by the WGS 84 Earth Gravitational Model (EGM) [*Defense Mapping Agency*, 1987] was used to calculate the deflection of the vertical and geoid height at each station using the Defense Mapping Agency program CLENQUENT [*Gleason*, 1985]. The form of the WGS 84 EGM is a spherical harmonic expansion of the gravitational potential; we used an expansion to degree and order 360. The gravitational coefficients are from *Rapp and Cruz* [1986] and R. Rapp (personal communication, 1988). The reference ellipsoid used to calculate the deflections of the vertical and the geoid - ellipsoid height separation was GRS 80 [*Defense Mapping Agency*, 1987], which is used in both the WGS 84 and NAD 83 reference systems. The deflections of the vertical vary in the network from 1.14"S to 0.14"N for ξ and from 1.40"E to 3.60"E for η. Utilizing the geoid - reference ellipsoid height separation and deflections of the vertical calculated from the long-wavelength part of the geopotential, the strain rate parameters were estimated using the DYNAP method; yielding $\gamma = 0.12 \pm 0.09$ μstrain/yr and $\beta = N17°E \pm 21°$ and an increase in the rms misfit by 2%. The strain rate results are similar to those obtained without corrections. The surface topography as well as the density distribution within the crust of the San Benito region is irregular, so the geoid representation used represents poorly the higher order features of the gravity field.

Because we expect significant short wavelength variations in the geoid, we have obtained from the Defense Mapping Agency geoid - reference ellipsoid separations and deflections of the vertical computed for network station positions from local gravity observations. The deflections of the vertical vary in the network from 6.00"S to 7.38"N for ξ and from 12.47"E to 6.38"W for η. The strain rate parameters utilizing these corrections are $\gamma = 0.19 \pm 0.09$ μstrain/yr and $\beta = N16°E \pm 13°$. The rms misfit decreases by 8%. This decrease is due to a lower misfit for the direction observations; the corrections do not change the rms misfit for distance observations. These results suggest that while it is desirable to correct for deflection of the vertical and for the separation between the geoid and the reference ellipsoid, these corrections are valuable only when based on local observations.

Acknowledgements. We thank Wayne Thatcher, Will Prescott, Jerry Eaton, and Jim Savage for assistance while JS was in Menlo Park; Richard Snay for the triangulation direction lists; Alice Drew, Kurt Feigl, and Bob King for their help in the use of DYNAP; and L. Decker and H. White of the Defense Mapping Agency for providing values for the deflection of the vertical and the geoid - reference ellipsoid separation. We also thank Tom Jordan for his constructive suggestions. Trilateration surveys were performed by the USGS Crustal Strain Project under the direction of Will Prescott. Support for part of this research was provided by the U.S. Geological Survey while JS was in Menlo Park. The portion of this research conducted at the Massachusetts Institute of Technology was supported by the National Aeronautics and Space Administration through a Graduate Student Research Fellowship (NGT-50103) to JS and through the Crustal Dynamics Project under grant NAG 5-814.

References

Anderson, W.L., Weighted triangulation adjustment, *U.S. Geol. Surv. Open File Rep.*, Computer Contribution Number 1, 52 pp., U.S. Geol. Surv. Computer Center Div., Washington, D.C., 1969.

Biot, M.A., Theory of folding of stratified viscoelastic media and its implications in tectonics and orogenesis, *Geol. Soc. Am. Bull.*, 72, 1592-1620, 1961.

Bird, P., and R.W. Rosenstock, Kinematics of present crust and mantle flow in southern California, *Geol. Soc. Am. Bull.*, 95, 946-957, 1984.

Bomford, G., *Geodesy*, 855 pp., Clarendon Press, Oxford, 1980.

Burford, R.O., Strain analysis across the San Andreas fault and Coast Ranges of California, Ph.D. thesis, 74 pp., Stanford University, Stanford, California, 1967.

Burford, R.O., and P.W. Harsh, Slip on the San Andreas fault in central California from alinement array surveys, *Bull. Seismol. Soc. Am.*, 70, 1233-1261, 1980.

Clark, M.M., K. Harms, J. Lienkaemper, D. Harwood, K. Lajoie, J. Matti, J. Perkins, M. Rymer, A. Sarna-Wojcicki, R. Sharp, J. Sims, J. Tinsley, and J. Ziony, Preliminary slip-rate table and map of late Quaternary faults of California, *U.S. Geol. Surv. Open File Rep.*, 84-106, 12 pp., 1984.

Clark, T.A., D. Gordon, W.E. Himwich, C. Ma, A. Mallama, and J.W. Ryan, Determination of relative site motions in the western United

TABLE 4. Summary of Deformation Rates within the Central Coast Ranges

Fault	Orientation	Geological Slip Rate, mm/yr	Geodetic Slip Rate, mm/yr
Right-lateral strike-slip faults			
Ortigalita	N35°W	0-2[a]	
San Andreas	N41°W	31-37[b]	32 ± 3[c]
Rinconada	N35°W	0-2[d]	2 ± 1[e]
San Simeon	N34°W	6-9[f]	
San Gregorio	N20°W	7-11[g]	0 ± 8[h]
Compression in the Coast Ranges			
East of San Andreas		2.2-5.5[i]	5.7 ± 2.7[j]
West of San Andreas		4.4-11[i]	6.1 ± 1.7[k,l]

[a] *Hart et al.* [1986].

[b] *Sieh and Jahns* [1984].

[c] *Savage and Burford* [1973]; *Thatcher* [1979a]; *Burford and Harsh* [1980]; *Lisowski and Prescott* [1981]; this study.

[d] *Hart et al.* [1986]; D. B. Slemmons, personal communication, 1987; E. W. Hart, personal communication, 1988.

[e] The line between Brush and Mulligan of the USGS Pajaro trilateration network has been measured five times between May 1978 and April 1983. If it is assumed that the average line length change is due to right-lateral slip on the King City fault, a northern extension of the Rinconada fault, then a slip rate of 2 ± 1 mm/yr is indicated.

[f] Rate of right-vertical oblique slip inferred from the offset of an ancient marine shoreline; a preferred slip rate of 6 mm/yr is given by *Clark et al.* [1984].

[g] Rate of right-lateral slip inferred from the offset of an ancient marine shoreline; a preferred slip rate of 7 mm/yr is given by *Clark et al.* [1984].

[h] From trilateration measurements made to the Farallon Islands between mid-1979 and late 1985 [*Prescott and Yu*, 1986].

[i] *Namson and Davis* [1988] estimated that 11 km of late Cenozoic shortening has occurred between the San Andreas fault and the Great Valley. The 22 km of shortening to the west of the San Andreas fault was computed by *Namson and Davis* [1988] from a solution that satisfies the observed structural relief. The range in rate estimates was obtained by assuming that shortening commenced between 5 and 2 m.y. ago.

[j] This study. The direction of maximum contraction is N16°E ± 13°.

[k] *Segall and Harris* [1986]; *Harris and Segall* [1987]. Average rates of change of line length from the San Luis trilateration network were used to invert for slip rate at depth on the San Andreas fault. In order to fit the trilateration measurements from this network, it was necessary to include a component of contraction normal to the trend of the San Andreas fault. The inversion results suggest a spatially uniform normal strain of -0.06 μstrain/yr. The net shortening rate across the network is 6.1 ± 1.7 mm/yr. This estimated compression, however, may be due to a systematic bias in the older trilateration data [J.C. Savage, personal communication, 1987].

[l] Two additional geodetic studies using historical triangulation data have been made west of the San Andreas fault. *Burford* [1967] analyzed triangulation data measured between 1930 and 1951 from two networks. One extends from Monterey Bay to the region where the San Andreas and Calaveras faults diverge (Figure 1), and one extends from Kettleman Hills near Parkfield west to San Luis Obisbo. Outside a zone close to the San Andreas fault, the direction of maximum shortening was estimated to be approximately N35°E. As part of a general study of the deformation in central California, *Thatcher* [1979a] examined triangulation data measured during the time interval 1944-1963 from the Salinias Valley network. Most of this network lies within the Salinian block located between the San Andreas and Rinconada faults. Examination of subregions suggest that the strains are poorly resolved, with the orientation of the inferred strain field not correlated with any known faults or tectonic trends. Because of the uncertainties in the results, neither of these studies were used for rate determinations.

[*Biot*, 1961; *Harland*, 1971]. At some point beyond this initial stage less work may be required to break new faults or build new folds, and the folding and faulting may commence elsewhere.

To explore the implications of the constraints provided by the San Andreas discrepancy vector we compare the predicted rate of ~5 mm/yr of fault-parallel and ~8 mm/yr of fault-normal deformation to the values given in Table 4. The rates of right-lateral slip on the Rinconada and San Gregorio faults estimated from geological observations are 0-2 and 6-9 mm/yr, respectively [*Clark et al.*, 1984]. These values compare with geodetically measured rates of 2 ± 1 mm/yr (Table 4) and 0 ± 8 mm/yr [*Prescott and Yu*, 1986], respectively. These results for the rate of slip on the San Gregorio fault are significantly smaller than some earlier estimates [*Minster and Jordan*, 1984; *Weldon and Humphreys*, 1986] but are more consistent with recent kinematic models [e.g., Model D, *Minster and Jordan*, 1987].

Geological and seismicity data suggest that northeast-southwest compression across the southern Coast Ranges may be localized to two regions, the 30-km-wide zone spanned by the San Benito network and a second zone to the west of the Rinconada fault [*Eaton*, 1984; *Dehlinger and Bolt*, 1987]. If shortening across the Coast Ranges is divided equally between these two regions, approximately 4 mm/yr of shortening should be occurring within the Diablo Range. If in contrast, the predicted 8 mm/yr [*Minster and Jordan*, 1987] is distributed uniformly across the 170-km-wide zone between the continental escarpment and the Great Valley, the shortening across the region spanned by the San Benito network would be approximately 1.4 mm/yr. The rate of deformation estimated from the San Benito study is most consistent with the first hypothesis, but the uncertainties in our calculated values do not allow the latter possibility to be ruled out.

Results from ongoing geodetic studies with stations in central California should provide additional constraints on the rate and distribution of slip within the Coast Ranges. The rate of slip estimated from VLBI measurements made at Vandenberg, Fort Ord, Presidio, and Point Reyes (Figure 6) could potentially be used to constrain the rate of deformation across the Coast Ranges. To distinguish between different models will require that uncertainties in the rate of slip, relative to a fixed reference, be 2-3 mm/yr or less. Our preliminary analysis of VLBI data, based on measurements between October 1982 and March 1987, indicate that only the rate of slip at the station Vandenberg meets this requirement [*Sauber et al.*, 1987; see also *Clark et al.*, 1987]. Additional measurements at all of the VLBI stations in central California have been made within the last year, and an analysis of these measurements will be the subject of a later paper. The rate of deformation estimated from measurements to the Farallon Islands (Figure 6) will be updated on the basis a trilateration survey of the network by the USGS in 1988; such information should provide a better constraint on the rate of slip on the San Gregorio fault.

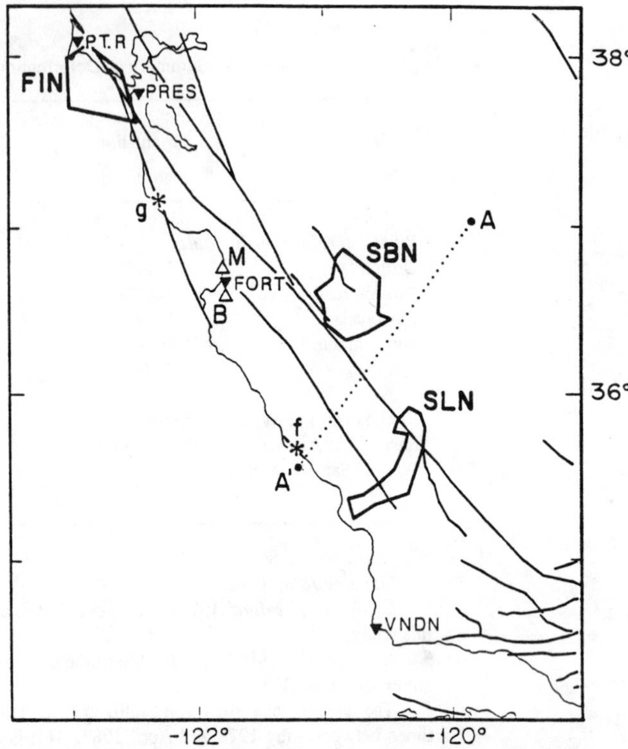

Figure 6. Reference figure for the geodetic and geological studies cited in Table 4. Brush (B) and Mulligan (M) are two stations in the USGS Pajaro trilateration network. At sites f and g, rates of fault slip were estimated geologically [*Clark et al.*, 1984] for the San Gregorio and Hosgri faults. The line A-A' is that for which *Namson and Davis* [1988] constructed their geological cross section (see Table 4). Fault traces are as given in Figure 1. Geodetic networks include FIN = Farallon Islands network, SBN = San Benito network, and SLN = San Luis network. VLBI stations include FORT = Fort Ord, PRES = Presidio, PT. R = Point Reyes, and VNDN = Vandenberg.

Summary

Triangulation and trilateration data from two geodetic networks located between the San Andreas fault and the Great Valley have been used to calculate shear strain rates in the Diablo Range and to estimate the slip rate along the Calaveras and Paicines faults in central California. The shear strain rates, $\dot{\gamma}_1$ and $\dot{\gamma}_2$, were estimated independently from angle changes using Prescott's method and from the simultaneous reduction for station position and strain parameters using the DYNAP method. On the basis of Prescott method, the average shear strain rate for the time period between 1962 and 1982 is 0.15 ± 0.08 μrad/yr, with the orientation of the most compressive strain (β) at N16°E ± 14°. Utilizing the DYNAP method with corrections for the deflection of the vertical and the geoid - reference ellipsoid separation computed on the basis of local gravity observations, $\dot{\gamma}$ = 0.19 ± 0.09 μrad/yr and β = N16°E ± 13°. At the 95% confidence level $\dot{\gamma}$ is not significantly greater than zero. The orientation of β, however, is similar to the orientation of maximum compressive strain indicated by the orientations of major fold structures in the region (N25°E). We infer that the measured strain is due to compression across the folds of this area; the average shear straining corresponds to a relative shortening rate of 5.7 ± 2.7 mm/yr.

The orientations of maximum principal stress inferred from wellbore breakouts and the azimuths of P axes determined from earthquake focal mechanisms within the Diablo Range and near the western edge of the Great Valley are similar to the orientation of maximum compressive strain implied by the trend of local fold structures. In contrast to the situation throughout most of the Coast Ranges in central California where fold axes have orientations approximately parallel to the San Andreas fault, within the Diablo Ranges between Hollister and Coalinga the trends of the fold axes are different and are thought to be controlled by reactivation of older structures. Given such structural control, the geodetic data reported here are consistent with transpression across the Coast Ranges.

For a zone within 10 km of the San Andreas fault, trilateration measurements on off-fault lines east of the San Andreas fault as well as lines that cross the San Andreas fault have been used to estimate the rate of slip along the Calaveras-Paicines fault and to document the gradual southward transition in the width of the zone accommodating right-lateral fault slip. South of Hollister the inferred rate of slip on the Calaveras-Paicines fault was found to be 10-12 mm/yr. The rate of slip on the Paicines fault near Bitter is ~4 mm/yr. Further to the south all of the right-lateral slip (at least across the 20-km-wide zone of measurements) occurs on the San Andreas fault.

To distinguish between different models that describe the distribution of strike-slip and compressive displacements within the Coast Ranges we examined data from regional geologic and geodetic studies and global plate models. Geological and seismicity data [*Eaton*, 1984; *Dehlinger and Bolt*, 1987], as well as our geodetic results, suggest that northeast-southwest compression across the southern Coast Regions may be localized to two regions, although uniform compression across the 170-km-wide zone between the continental escarpment and the Great Valley cannot be ruled out.

Appendix. Details of Data Reduction for the San Benito Network

Accuracy of the Triangulation and Trilateration Measurements

Triangulation is a measurement system consisting of joined or overlapping triangles of angular observations. During a single session observations of direction are made from a particular mark to several other marks that are located within a few tens of kilometers from the mark occupied. An angle observation is determined by differencing two direction measurements. Seventy of the 72 directions utilized in this study were second-order observations, with a priori uncertainty estimated to be σ_d = 3.4 μrad [*Federal Geodetic Control Committee*, 1984], and two were third-order observations (σ_d =5.8 μrad). The expected uncertainty in a second-order angle measurement (σ_a) is $\sqrt{2}\,\sigma_d$, or about 4.8 μrad. The order of the triangulation denotes the measurement precision, which is determined by survey procedures and is reflected in the degree to which internal checks of the data are satisfied [*Thatcher*, 1979b]. The principal internal check is the triangle closure requirement that the angles within each triangle sum to 180° plus the known excess due to the Earth's sphericity. Triangle closures for these data indicate that the standard deviation of a single angle is approximately 5.3 μrad, very close to the expected value.

As practiced by the USGS, trilateration consists of distance measurements among a network of stations. The distances between geodetic monuments were measured in this study with a Geodolite, a precise electro-optical distance-measuring instrument. Regional trilateration measurements made by the USGS on line lengths of 10-40 km have a precision of approximately 0.2-0.3 ppm [*Savage and Prescott*, 1973]. After corrections for refractivity and instrument and reflector height, a line length corresponds to the distance between two geodetic monuments.

Assumptions Made in the Estimation of Horizontal Shear Parameters

The full three-dimensional strain tensor includes horizontal and vertical shear components, dilatation, and vector rotation. Since height changes were not directly measured in either the 1962 or 1982 survey, we can estimate only the horizontal components of the strain tensor. Here we assume that vertical changes in station height are negligible. Since no large earthquakes occurred within the region spanned by the San Benito network between 1932-1982 [*Engdahl and Rinehart*, 1989], the estimated rate of uplift on the folds of the Diablo Range is only 1-3 mm/yr [*Zepeda et al.*, 1987], or 2-6 cm over the 20-year time period of this study. A 6-cm change in the height of Tum (Figure 2) between 1962 and 1982, for example, will cause a 0.1 ppm change in the 20-km-length line between Tum and Bonito (calculated from equation 1.72 of *Bomford* [1980]). Because of ground-water-induced subsidence, larger vertical changes may have occurred at the stations Stubble, located in the Great Valley, and Panoche, located in Panoche Valley. Further, astronomic azimuth measurements are made only at Hepsedam. Without a reliable external or conventionally adopted internal reference direction, we cannot estimate rotation of the network about a vertical axis. Additionally, since length measurements were made only in the 1982 survey, surface dilatation can not be estimated. We are thus able to estimate only the rate of change of horizontal shear components $\dot{\gamma}_1$ and $\dot{\gamma}_2$. If the fold structures of the region are deforming as the result of simple uniaxial compression, block rotation is not expected and dilatation is assumed to be uniform.

The two modes of deformation suggested from the geological structures of the region (see text) can be used to interpret the values of $\dot{\gamma}_1$ and $\dot{\gamma}_2$. Right-lateral strike-slip motion at the orientation of the San Andreas fault (N41°W) would be seen primarily as $\dot{\gamma}_1 > 0$. North-south compression is also consistent with $\dot{\gamma}_1 > 0$. If the orientation of compressional strain is northeast-southwest, as is predicted from the orientation of the folds within the southern Coast Ranges, then $\dot{\gamma}_2 < 0$.

Prescott's Method

Angle changes are the observations used to estimate $\dot{\gamma}_1$ and $\dot{\gamma}_2$ in the extended version of Frank's method [*Prescott*, 1976]. The fundamental equation of the technique,

$$\Delta\Phi_i = (t_i - t_o)\left(A_i^1\,\dot{\gamma}_1 + A_i^2\,\dot{\gamma}_2\right) \tag{A.1}$$

where

$$A_i^1 = \left[(\sin\theta_{i2} - \sin\theta_{i1})\,\dot{\gamma}_1\right]/2$$

$$A_i^2 = \left[(\cos\theta_{i2} - \cos\theta_{i1})\,\dot{\gamma}_2\right]/2$$

relates an observed angular change $\Delta\Phi_i$ for the time interval $t_i - t_o$ to the parameters $\dot{\gamma}_1$ and $\dot{\gamma}_2$. Here θ_{i1} and θ_{i2} represent the azimuths (clockwise from north) of the initial and terminal sides of the angle. To derive angles the 1982 data were used to adjust for station position. Rather than making an adjustment on a three-dimensional surface, the distances were projected onto the Clark 1866 reference ellipsoid (NAD27 geodetic system, *Defense Mapping Agency* [1987] and an adjustment was made employing a variation of coordinates method [*Anderson*, 1969]. Azimuths for each station-to-station pair were then determined for comparison to the 1962 observations. In this approach we are comparing angles measured on the Earth's surface to angles determined from a network adjustment on the reference ellipsoid. The classical solution to the adjustment of geodetic data derived from different measurement techniques has involved making corrections to the direction and distance observations such that the measurements are then given on a common reference ellipsoid [*Bomford*, 1980; *Vanicek and Krakiwsky*, 1986]. In employing the DYNAP method an alternative approach is used.

DYNAP Method

In the DYNAP (DYNamic Adjustment Program) technique the directions measured in 1962 and the distances measured in 1982 are used simultaneously to solve for both crustal motion parameters and positional coordinates of the geodetic marks for a specified reference time via weighted least squares [*Snay*, 1986; *Drew and Snay*, 1988]. A two-dimensional adjustment was carried out by holding the station elevations fixed. In this approach the direction observations are corrected for the deflection of the vertical, and the separation of the geoid and a reference ellipsoid are used to correct the distance observations. This technique is based on the assumption that the velocity field is linear in space and constant over the time interval of interest. The time dependent station positions can be written as

States using Mark III very long baseline interferometry, *J. Geophys. Res.*, 92, 12,741-12,750, 1987.

Coleman, R., and K. Lambeck, Crustal motion in southeastern Australia: Is there geodetic evidence for it?, *Aust. J. Geod. Photo. Surv.*, 39, 1-26, 1983.

Crouch, J.K., S.B. Bachman, and J.T. Shay, Post-Miocene compressional tectonics along the central California margin, in *Tectonics and Sedimentation along the California Margin*, edited by J.K. Crouch and S.B. Bachman, Pac. Sect. Soc. Econ. Paleontol. Mineral., 38, 37-54, 1984.

Davis, T.L., A structural outline of the San Emigdo Mountains, in *Geologic Transect Across the Western Transverse Ranges*, edited by T.L. Davis and J. Namson, Pac. Sect. Soc. Econ. Paleontol. Mineral. Guidebook, pp. 23-32, 1986.

Defense Mapping Agency, Department of Defense World Geodetic System 1984: Its definition and relationships with local geodetic systems, *DMA Tech. Rep.*, 8350.2, 120 pp., 1987.

Dehlinger, P., and B.A. Bolt, Earthquakes and associated tectonics in a part of coastal central California, *Bull. Seismol. Soc. Am*, 77, 2056-2073, 1987.

DeMets, C., R.G. Gordon, S. Stein, and D.F. Argus, A revised estimate of Pacific-North American motion and implications for western North America plate boundary zone tectonics, *Geophys. Res. Lett.*, 14, 911-914, 1987.

Dibblee, T.W., Regional geology of the central Diablo Range between Hollister and New Idria, in *Field Trip Guidebook for the Geological Society of America Cordilleran Section Meeting*, edited by T.H. Nilsen and T.W. Dibblee, pp. 6-16, Geol. Soc. Am., Boulder, Colo., 1979.

Drew, A., and R. Snay, DYNAP: Software for estimating crustal deformation from geodetic data (abstract), *Eos Trans. AGU*, 69, 325, 1988.

Eaton, J.P., Focal mechanisms of near-shore earthquakes between Santa Barbara and Monterey, California, *U.S. Geol. Surv. Open File Rep.*, 84-477, 13 pp., 1984.

Eaton, J.P., Regional seismic background of the May 2, 1983 Coalinga earthquake, in *Proceedings of Workshop XXVII, Mechanics of the May 2, 1983 Coalinga Earthquake*, edited by M.J. Rymer and W.L. Ellsworth, *U.S. Geol. Surv. Open File Rep.*, 85-44, pp. 44-60, 1985.

Ellsworth, W.L., Bear Valley, California, earthquake sequence of February-March 1972, *Bull. Seismol. Soc. Am.*, 65, 483-506, 1975.

Engdahl, E.R., and W.A. Rinehart, Seismicity map of North America, in *Neotectonics of North America*, edited by D.B. Slemmons, Geol. Soc. Amer., Boulder, Colo., in press, 1989.

Federal Geodetic Control Committee, Standards and Specifications for Geodetic Control Networks, Nat. Ocean. Atmos. Admin., U.S. Dept. Commerce, Rockville, Md., 1984.

Frank, F.C., Deduction of earth strains from survey data, *Bull. Seismol. Soc. Am.*, 56, 35-42, 1966.

Gawthrop, W.J., Seismicity and tectonics of the central California coastal region, M.S. thesis, University of Colorado, Boulder, 76 pp., 1977.

Gleason, D.M., Partial sum of Legendre series via Clenshaw summation, *Manuscripta Geodetica*, 10, 115-130, 1985.

Hamilton, D.H., Characterization of the San Gregorio-Hosgri fault system, coastal central California (abstract), *Geol. Soc. Am. Abstr. Programs*, 19, 385, 1987.

Harding, T.P., Tectonic significance and hydrocarbon trapping consequences of sequential folding synchronous with San Andreas faulting, San Joaquin Valley, California, *Am. Assoc. Petro. Geol. Bull.*, 60, 356-378, 1976.

Harland, W.B., Tectonic transpression in Caledonian Spitsbergen, *Geol. Mag.*, 108, 27-42, 1971.

Harms, K.K., J.W. Harden, and M.M. Clark, Use of quantified soil development to determine slip rates on the Paicines fault, northern California (abstract), *Geol. Soc. Am. Abstr. Programs*, 19, 387, 1987.

Harris, R.A., and P. Segall, Detection of a locked zone at depth on the Parkfield, California, segment of the San Andreas fault, *J. Geophys. Res.*, 92, 7945-7962, 1987.

Harsh, P.W., and N. Pavoni, Slip on the Paicines fault, *Bull. Seismol. Soc. Am.*, 68, 1191-1193, 1978.

Hart, E.W., W.A. Bryant, M.W. Manson, and J.E. Kahle, Summary report: Fault evaluation program 1984-1985, south Coast Ranges region and other areas, *Calif. Div. Mines Geol. Open File Rep.*, 86-3SF, 1986.

Heiskanen, W.A., and H. Moritz, *Physical Geology*, 364 pp., Freeman, San Francisco, 1967.

Horns, D.M., C.Y. Wang, and Y. Shi, Investigation of the uplift along the San Andreas fault zone (abstract), *Eos Trans. AGU*, 66, 1093, 1985.

Jennings, C.W., Fault map of California, *Calif. Geol. Data Map Ser.*, map 1, Calif. Div. of Mines and Geol., Sacramento, 1975.

Lisowski, M., and W.H. Prescott, Short-range distance measurements along the San Andreas fault system in central California, 1975 to 1979, *Bull. Seismol. Soc. Am.*, 71, 1607-1624, 1981.

Lockwood, J.P., and J.G. Moore, Regional deformation of the Sierra Nevada, California, on conjugate microfault sets, *J. Geophys. Res.*, 92, 2747-2766, 1987.

Malvern, L.E., *Introduction to the Mechanics of a Continuous Medium*, 713 pp., Prentice-Hall, Englewood Cliffs, N.J., 1969.

Matsu'ura, M., D.D. Jackson, and A. Cheng, Dislocation model for aseismic crustal deformation at Hollister, California, *J. Geophys. Res.*, 91, 12,661-12,674, 1986.

Minster, J.B., and T.H. Jordan, Vector constraints on Quaternary deformation of the western United States east and west of the San Andreas fault, in *Tectonics and Sedimentation along the California Margin*, edited by J.K. Crouch and S.B. Bachman, Pac. Sect. Soc. Econ. Paleontol. Mineral., 38, 1-16, 1984.

Minster, J.B., and T.H. Jordan, Vector constraints on western U.S. deformation from space geodesy, neotectonics, and plate motions, *J. Geophys. Res.*, 92, 4798-4804, 1987.

Mount, V.S., and J. Suppe, State of stress near the San Andreas fault: Implications for wrench tectonics, *Geology*, 15, 1143-1146, 1987.

Namson, J.S., and T.L. Davis, Seismically active fold and thrust belt in the San Joaquin Valley, central California, *Geol. Soc. Am. Bull.*, 100, 257-273, 1988.

Page, B.M., The southern Coast Ranges, in *The Geotectonic Development of California*, edited by W. G. Ernst, pp. 329-417, Prentice Hall, Englewood Cliffs, N.J., 1981.

Page, B.M., Geologic background of the Coalinga earthquake of May 2, 1983, in *Proceedings of Workshop XXVII, Mechanics of the May 2, 1983 Coalinga Earthquake*, edited by M.J. Rymer and W.L. Ellsworth, *U.S. Geol. Surv. Open File Rep.*, 85-44, pp. 4-9, 1985.

Page, B.M., and D.C. Engebretson, Correlation between the geologic record and computed plate motions for central California, *Tectonics*, 3, 133-155, 1984.

Prescott, W.H., An extension of Frank's method for obtaining crustal shear strains from survey data, *Bull. Seismol. Soc. Am.*, 66, 1847-1853, 1976.

Prescott, W.H., and M. Lisowski, Strain accumulation along the San Andreas fault system east of San Francisco Bay, California, *Tectonophysics*, 97, 41-56, 1983.

Prescott, W.H., and S.Yu, Geodetic measurement of horizontal deformation in the northern San Francisco Bay region, California, *J. Geophys. Res.*, 91, 7475-7484, 1986.

Rapp, R.H., and J.Y. Cruz, The representation of the Earth's gravitational potential in a spherical harmonic expansion to degree 250, *Air Force Geophys. Lab. Tech. Rep.*, 86-0191, 64 pp., 1986.

Raymond, L.A., Tesla-Ortigalita fault, Coast Range thrust fault, and

Franciscan metamorphism, northeastern Diablo Range, California, *Geol. Soc. Am. Bull., 84*, 3547-3562, 1973.

Sauber, J., T.H. Jordan, G.C. Beroza, T.A. Clark, and M. Lisowski, Constraints on North American-Pacific plate boundary deformation in central California from VLBI and ground-based geodetic data (abstract), in *Programs and Abstracts, The Impact of VLBI on Astrophysics and Geophysics*, Inter. Astron. Un. Symp. No. 129, p. 7.1, 1987.

Saucier, F., and E. Humphreys, Finite element kinematic model of southern California (abstract), *Eos Trans. AGU, 69*, 331, 1988.

Savage, J.C., Strain accumulation in western United States, *Ann. Rev. Earth Planet. Sci., 11*, 11-43, 1983.

Savage, J.C., and R.O. Burford, Geodetic determination of relative plate motion in central California, *J. Geophys. Res., 78*, 832-845, 1973.

Savage, J.C., and W.H. Prescott, Precision of Geodolite distance measurements for determining fault movements, *J. Geophys. Res., 78*, 6001-6008, 1973.

Segall, P., and R. Harris, Slip deficit on the San Andreas fault at Parkfield, California, as revealed by inversion of geodetic data, *Science, 233*, 1409-1413, 1986.

Segall, P., and D.D. Pollard, The mechanics of discontinuous faults, *J. Geophys. Res., 85*, 4337-4350, 1980.

Sieh, K.E., and R.H. Jahns, Holocene activity of the San Andreas fault at Wallace Creek, California, *Geol. Soc. Am. Bull., 95*, 883-896, 1984.

Slemmons, D.B., Capable faults and tectonically active folds of the California central Coast Ranges (abstract), *Geol. Soc. Am. Abstr. Programs, 19*, 452, 1987.

Snay, R.A., Horizontal deformation in New York and Connecticut: Examining contradictory results from the geodetic evidence, *J. Geophys. Res., 91*, 12695-12702, 1986.

Springer, J.E., Stress orientations from wellbore breakouts in the Coalinga region, *Tectonics, 6*, 667-676, 1987.

Stein, R.S., Evidence for surface folding and subsurface fault slip from geodetic elevation changes associated with 1983 Coalinga, California earthquake, in *Proceedings of Workshop XXVII, Mechanics of the May 2, 1983 Coalinga Earthquake*, edited by M.J. Rymer and W.L. Ellsworth, *U.S. Geol. Surv. Open File Rep.*, 85-44, pp. 225-253, 1985.

Stein, R.S., and G.C.P. King, Seismic potential revealed by surface folding: 1983 Coalinga, California, earthquake, *Science, 224*, 869-872, 1984.

Thatcher, W., Systematic inversion of geodetic data in central California, *J. Geophys. Res.*, 84, 2283-2295, 1979a.

Thatcher, W., Horizontal crustal deformation from historic geodetic measurements in southern California, *J. Geophys. Res., 84*, 2351-2370, 1979b.

Thatcher, W., Geodetic measurement of active-tectonic processes, in *Active Tectonics*, pp. 155-163, National Academy Press, Washington, D.C., 1986.

Vanicek, P., and E. Krakiwsky, *Geodesy: The Concepts*, 697 pp., Elsevier, New York, 1986.

Weldon, R., and E. Humphreys, A kinematic model of southern California, *Tectonics, 5*, 33-48, 1986.

Weldon, R. and E. Humphreys, Plate model constraints on the deformation of coastal southern California north of the Transverse Ranges (abstract), *Geol. Soc. Am. Abstr. Programs, 19*, 462, 1987.

Zepeda, R.L., E.A. Keller, and T.K. Rockwell, Soil chronosequence at Wheeler Ridge, southern San Joaquin Valley, California (abstract), *Geol. Soc. Am. Abstr. Programs, 19*, 466, 1987.

Zoback, M.D., M.L. Zoback, V.S. Mount, J. Suppe, J.P. Eaton, J.H. Healy, D. Oppenheimer, P. Reasenberg, L. Jones, C.B. Raleigh, I.G. Wong, O. Scotti, and C. Wentworth, New evidence on the state of stress of the San Andreas fault sytem, *Science, 238*, 1105-1111, 1987.

EARTHQUAKES' IMPACT ON CHANGES IN HEIGHT

I. Joó

College for Surveying and County Planning of University of Forestry and Timber Industry H-8000 Székesfehérvár, Pirosalma u. 1–3.

Abstract. The country of Hungary is sufficiently well surveyed from the point of view of vertical crustal movements. The area north of Lake Balaton rising at the rate of 0.2 mm/a. There was an earthquake of magnitude of 5.6 (Richter scale) in 1985 close to Berhida and Peremarton. A firstorder levelling line crosses this area of variable terrain. A re-measurement of the line was carried out after the earthquake. The changes in height caused by the earthquake, the relative vertical velocities, horizontal gradients of velocities and the correlation between the tectonic structure and changes in height is discussed in this paper.

Introduction

The Carpathian Basin, wich includes Hungary, is situated at the centre of Europe. The Basin is surrounded by the zone of the Carpathian Mountains. The last main phase of uplift of the Carpathian Mountains was in middle Miocene and Pliocene 100-130 million years ago, like the Eastern Alps and its rolling up ended about 15-50 million years ago Miocene. (Horváth 1984,) (Fig.1.)

Development of the Hungarian Central Range Midmountains ended in the Cretaceous, but a depression, the Great Hungarian Plain, formed since the Oligocene (Balla Z. 1984,).

A significant part of the Carpathian Basin was formerly covered by the Central Paratethys Sea. The sea gradually retreated as a result of sedimentation which followed the erosion of the mountains and other tectonic effects. In places, the sediment is 1000 to 6000 m thick.

Copyright 1989 by
International Union of Geodesy and Geophysics and American Geophysical Union.

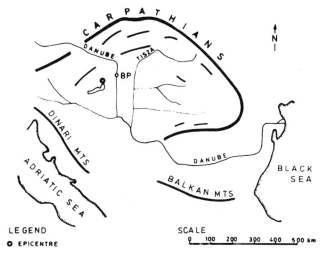

Fig. 1. The Carpathian Mountains

Volcanism and uplift modified the Carpathian Basin in the Oligocene. Subsidence started in the early Miocene, and accelerated in the middle to late Miocene with decreasing volcanic activity. During Tertiary and Quaternary a system of faults developed. (Pospisil-Vass, 1984,)

The area we investigated is located at the northern corner of Lake Balaton at the intersection of the Great Hungarian Plain and Transdanubian Midmountains. In this area there was an earthquake of magnitude of 5.6 on the Richter scale on August 15, 1985 with numerous aftershocks. The epicentre of the shock was located by seismologists near Berhida-Peremarton-latitude $47°05'$ longitude (east) $18°07'$.

The hypocentre was in depth of about 2.0-2.8 km (Fig.2.) (Szeidovitz, 1987).

We investigated the vertical motions caused by the earthquake using repeated levellings. A first-order levelling line

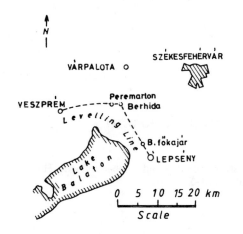

Fig. 2. Area of the earthquake occured in August 1985

measured in 1976 crosses the area of the earthquake, facilitating geodetic investigation of vertical movement.(Fig.3.) After the earthquake a levelling line 39 km long containing 53 bench marks was re-measured in October through December, 1985 between Veszprém and Lepsény.

At Lepsény there is a bench mark with a deep foundation, and there are eight special bench marks along the levelling line. One point was instable. Five points were determined in 1985, so 47 bench marks could be used for the purposes of investigation (Joó et al. 1986).

The western end of the line is on the Veszprém plateau which is a part of Southern-Bakony. The levelling line crosses the Basin of Litér, the Fault of Litér, the Vörösberény-Királyszent-

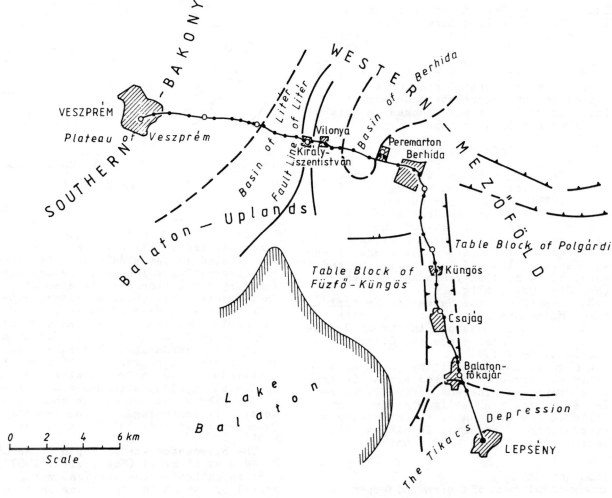

Fig. 3. The levelling line and the geological situation

istván tectonic unit and the Basin of Berhida, and twice intersects the stream Séd. After Berhida the line goes to South-East-South and ends at a principal bench mark at Lepsény in Tikacs depression. The Tikacs and Berhida Basins are typical depressions of Neopleistocene and Holocene age.

Veszprém plateau is a karst region formed of dolomite, in some places with loess cover.

In the Balaton Uplands an overlaying stratum about 1500 m in thichness lies on Paleozoic fill and Permian limestone which is cut by fault lines parallel with and perpendicular to the strike direction. The Basin of Litér trends North to East and is bordered by two fault lines.

The Vörösberény-Királyszentistván tectonic unit is cut by 12 smaller transverse fractures. Our area is on the north part of the plateau, around Királyszentistván.

The area of Mezőföld was originally covered by the Pannonion sea. It is now a plateau inclined to the southeast, covered by very rich microtechtonic carvings. The structure of the basement can well be detected on the surface. (Ádám-Marosi, 1959)

Methods of investigation

The investigation's main elements – in addition to collecting and analysing morpho-tectonic data – were the following:
a.) detailed reconnaissance the line of investigation
b.) collecting and analysing geodetic data.
During the reconnaissance the line of investigation
- the foundation and position of bench marks were checked
- supplementary information was collected on the earthquake
- the position of bench marks were compared to geologic information for determining what geologic unit they belong to.

All this knowledge was useful in the interpretation of results, too.

Both measurements (1976 and 1985) were carried out according to the rules of first order precise levelling. Raw data were used for investigation, with only the corrections of levelling staff comparation and height refraction applied to the data (Joó, 1987).

Reliable levelling data used for this investigation is characterised as follows:

a.) In Hungary the difference of height differences between forward and backward runnings should not exceed the following:
$$\Delta h = \pm 1.2 \sqrt{L} \text{ mm}$$
b.) The distance between adjacent bench marks should be 0.7-1.5 km. In our case it was 0.66 km.
c.) Besides conventional bench marks, there are points of deep foundation at 4-5 km intervals (max.depth 6 m).
d.) At 50-70 km intervals there are twin principal bench marks of deep foundation built (max.depth of 16 m).
e.) The a posteriori mean square error from the first measurement was
$$m_I = \pm 0.34 \text{ mm}/\sqrt{km}$$
and from the second measurement:
$$m_{II} = \pm 0.29 \text{ mm}/\sqrt{km}.$$
f.) The mean of differences of the height differences of forward and backward runnings is 0.44 mm, while the greatest difference was 1.04 mm.

Following values were derived from data of repeated levellings for every levelling section:

$$\Delta H_i = \Delta h_2 - \Delta h_1 \qquad (1)$$

$$\sum_V^i \Delta H \qquad (2)$$

$$\Delta V_i = \frac{\Delta H_i}{\Delta T} \qquad (3)$$

$$\text{hor.grad.} = \frac{\Delta V_i}{L_i} \varrho'' \quad ("/a) \qquad (4)$$

where Δh_1 and Δh_2 are differences in height derived from the first and second levellings, respectively,

ΔH is the change in height between two levellings
ΔV is the relative velocity of the section (mm/a)
ΔT the time elapsed between the two measurements (in this case it is 9 years)
hor.grad. is the horizontal gradient of the vertical velocity of the section in "/a.

The results of the investigation are compiled both in tables and in a graphic from.

It should be, however, noted that derived data are relative ones. This fact means that the relative movement of a point can be interpreted as compared to the adjacent points; or movements rela-

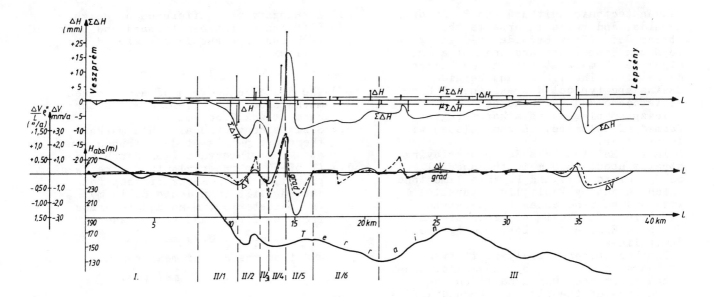

Fig. 4. Main results of investigation

tive to Veszprém can be obtained by means of summing up the changes.

On the basis of the measurements no closing error could be derived, therefore the reliability of the derived data can be estimated with limitations. For this reason mean square errors were introduced for the data of ΔH, $\Sigma\Delta H$, ΔV and hor.grad. using a posteriori standard error corresponding to the error propagation.

Main results of the investigation

The derived results are presented in Fig.4. where the derived ΔH values are discrete quantities. The $\Sigma\Delta H$ (upper, continuous) line is result of continuous summing up of ΔH values from Veszprém. Below this line the relative velocities (ΔV) can be seen. The values of horizontal gradient (dashed line) and the terrain profile are shown in the figure, too.

Roman figures represent geological units (see Table 1. last column) (Joó et al, 1986).

The most charachteristic movements could be detected on basis of Fig.4. at the section between the Litér-area and the basin of Berhida.

An interpretation of the ΔH values shown in Fig.4. has been given previously (see eq. 1). $\Sigma\Delta H$ values show relative movements of the end of the section nearer to Lepsény related to the starting point Veszprém.

ΔV values represent the relative velocity of a given section as a sum of apparent continuous movements in 9 year between the measurements, although it is most likely that the greater part of the movements occurred in a shorter time due to the earthquake.

Reliability of derived data

As the investigation was not carried out in a closed polygon, there was no possibility to find the closing error. For this reason the reliability of derived quantities was to be carefully estimated.

A posteriori standard errors derived from the first and the second measurements are

$$\mu_I = \pm\ 0.34\ \text{mm}/\sqrt{\text{km}}$$
$$\mu_{II} = \pm\ 0.29\ \text{mm}/\sqrt{\text{km}}.$$

The values of standard errors show that the rigorous instructions for first order precise levellings were observed.

In the middle part of the line, where movements of primary importance have occurred, (see section II/1 to II/6 in Fig.4.) there are two bench-marks belonging to each morphotectonic units. Trends of the derived characteristics for these bench-marks are the same in all cases and this fact shows that the derived characteristics are realistic ones.

Based on the propagation of errors, and using the mentioned a posteriori standard errors, the standard deviation of the derived data were computed.

The reliability of the computed ΔH is obtained considering the data of measurements as independent values as:

$$\mu_{\Delta H} = \pm \sqrt{\mu_I^2 + \mu_{II}^2} \quad (5)$$

Using the previous data the reliability of the derived ΔH values is

$$\mu_{\Delta H} = \pm\ 0.447\ \text{mm}/\sqrt{\text{km}}$$

i.e. approximately ± 0.45 mm/$\sqrt{\text{km}}$.

For computing the standard error $\mu_{\Delta H_i}$ of the derives concrete ΔH_i values for a levelling section the real length of the given section must be taken into consideration.
Therefore

$$\mu_{\Delta H_i} = \pm\ \mu_{\Delta H}\sqrt{L_i} \quad (6)$$

where L_i is the length of section i.

Data computed according to eq 6. are shown in Table 2.

Using the table, standard deviations $\mu_{\Delta H_i}$ can be computed for all ΔH_i values.

In this investigation not only concrete ΔH_i values for individual sections, but continuous sums from the starting point Veszprém were computed, too, therefore it is practical to compute the standard deviations for these $\Sigma \Delta H_i$ values using eq.6.

$$\mu_{\Sigma \Delta H_i} = \pm\ \mu_{\Delta H} \sqrt{\Sigma_V^i L} \quad (7)$$

Using ± 0.447 mm/$\sqrt{\text{km}}$ for $\mu_{\Delta H}$, Table 3. shows the values of the standard error for every 5 km distance ($\Sigma_V^i L$) counted from Veszprém.

The standard errors $\mu_{\Sigma \Delta H}$ for $\Sigma \Delta H$ values of each characteristic tectonic unit were computed using eq 7. (see Table 4.)

The standard error of horizontal gradients can be computed by using the error propagation:

$$\mu_{h.grad.} = \pm\ \mu_{\Delta V_i}\ \frac{\varrho''}{L_i} \quad (8)$$

For L=1 km, $\mu_{h.grad.}$ is $\pm 0.2\ \mu_{\Delta V_i}$

As the computed $\mu_{\Delta V}$ values are in the range of $\pm(0.02-0.08)$ mm/a and values of L_i are between 0.1 and 2.6 km, values of $\mu_{h.grad.}$ will be in the interval of $\pm(0.006-0.041)$ "/a.

Standard errors have been computed so for each ΔH_i, $\Sigma \Delta H_i$, ΔV and hor.grad. For $\mu_{\Sigma \Delta H}$, intervals $+|\mu_{\Sigma \Delta H}|$ and $-|\mu_{\Sigma \Delta H}|$ were indicated in Fig.4 (dashed line) to enable a comparison of the $\Sigma \Delta H$ values graphically. $\Sigma \Delta H$ values exceed this interval between the two bordering lines apart from the line between the Veszprém an Litér area - which is about 8 km in length.

Interpretation of results

Figure 4. has been already dealt with, it contains the most important features derived for the investigated line. In this chapter the most important results of investigation are discussed in detail. Data are given in each case.

- Small and not characteristic changes in height are derived for the Plateau of Veszprém. This is expected because the starting point of the investigated line is situated is this very geological unit.

- Change in height $\Sigma \Delta H$ is $-(6.16 \pm 2.79)$ mm between the two end points of the line (Veszprém-Lepsény).

- Most characteristic changes occurred at the section between area of Litér (II/1) and the basin of Berhida (II/5), and at the depression of Tikacs. They are as follows:

Area of Litér (II/1):

$$\Delta H = -(7.35 \pm 0.35)\ \text{mm}$$
$$\Sigma \Delta H = -(14.16 \pm 1.48)\ \text{mm}$$
$$\Delta V = -(0.82 \pm 0.04)\ \text{mm/a}$$
$$\text{hor.grad.} = -(0.28 \pm 0.014)\ ''/a.$$

Mogyorós-hill (Királyszentistván II/2):

$$\Delta H = (3.68 \pm 0.28)\ \text{mm}$$
$$\Sigma \Delta H = -(9.14 \pm 1.55)\ \text{mm}$$
$$\Delta V = (0.41 \pm 0.03)\ \text{mm/a}$$
$$\text{hor.grad.} = (0.21 \pm 0.015)\ ''/a.$$

Séd-stream (II/3):

$$\Delta H = -(6.68 \pm 0.20)\ \text{mm}$$
$$\Sigma \Delta H = -(19.37 \pm 1.62)\ \text{mm}$$
$$\Delta V = -(0.74 \pm 0.02)\ \text{mm/a}$$
$$\text{hor.grad.} = -(0.76 \pm 0.021)\ ''/a.$$

Tündér-hill (Vilonya II/4):

$$\Delta H = (23.68 \pm 0.28)\ \text{mm}$$
$$\Sigma \Delta H = (16.45 \pm 1.69)\ \text{mm}$$
$$\Delta V = (2.63 \pm 0.03)\ \text{mm/a}$$
$$\text{hor.grad.} = (1.36 \pm 0.015)\ ''/a.$$

Basin of Berhida (II/5):

$$\Delta H = -(26.26 \pm 0.40)\ \text{mm}$$
$$\Sigma \Delta H = -(9.81 \pm 1.74)\ \text{mm}$$
$$\Delta V = -(2.92 \pm 0.04)\ \text{mm/a}$$
$$\text{hor.grad.} = -(0.75 \pm 0.010)\ ''/a.$$

TABLE 1. Main Results of Geodetic Investigation

Number of Point	L (km)	ΣL (km)	ΔH (mm)	ΣΔH (mm)	ΔV (mm/year)	ΣΔV (mm/year)	ΔV/L ("/km)	Remarks
010								
151	0.3	0.3	+ 0.32	+ 0.32	0.04	0.04	0.02	
152	0.4	0.7	− 1.03	− 0.71	− 0.11	− 0.07	− 0.06	
153	0.7	1.4	+ 0.50	− 0.21	0.05	− 0.02	0.02	Plateau of
011	2.6	4.0	+ 0.04	− 0.17	0.01	− 0.01	∅	Veszprém
156	0.1	4.1	+ 0.49	+ 0.32	0.05	0.04	0.11	I.
157	0.6	4.7	− 0.70	− 0.38	− 0.08	− 0.04	− 0.03	
159	2.1	6.8	− 0.52	− 0.90	− 0.06	− 0.10	− 0.03	
160	0.6	7.4	+ 0.88	− 0.02	0.10	0.01	0.03	
161	0.7	8.1	− 0.28	− 0.30	− 0.03	− 0.03	− 0.01	
012	0.1	8.2	+ 0.20	− 0.10	0.02	− 0.01	0.04	
163	2.1	10.3	− 4.71	− 4.81	− 0.52	− 0.53	− 0.05	Basin of
164	0.6	10.9	− 7.35	−14.16	− 0.82	− 1.35	− 0.28	Litér (II/1.)
165	0.7	11.6	− 0.66	−12.82	− 0.07	− 1.42	− 0.02	Mogyorós hill
166	0.4	12.0	+ 3.68	− 9.14	0.41	− 1.01	0.21	(Királyszent-
013	0.1	12.1	+ 2.43	− 6.71	0.27	− 0.74	0.56	istván (II/2.)
016								
185	0.7	27.3	+ 0.31	− 4.82	0.03	− 0.54	0.01	
186	0.6	27.9	+ 1.84	− 2.98	0.20	− 0.33	0.07	
187	0.4	28.3	− 0.41	− 3.39	− 0.05	− 0.38	− 0.02	
188	0.6	28.9	+ 0.20	− 3.19	0.02	− 0.35	0.01	(III/1)
189	1.1	30.0	+ 1.08	− 2.11	0.12	− 0.24	0.02	
190	0.4	30.4	+ 0.99	− 1.12	0.11	− 0.12	0.06	
017	0.3	30.7	+ 0.14	− 0.98	0.02	− 0.11	0.01	
191	0.6	31.3	− 0.56	− 1.54	− 0.06	− 0.17	− 0.02	
193	1.5	32.8	+ 0.34	− 2.80	0.48	0.31	0.07	
194	1.0	33.8	+ 2.84	+ 0.04	− 0.32	∅	− 0.06	(III/2)
195	0.5	34.3	− 3.35	− 3.39	− 0.37	− 0.38	− 0.15	
196	0.6	34.9	+ 3.94	+ 0.55	0.44	0.06	0.15	
018	0.1	35.0	+ 1.22	+ 1.77	0.14	0.20	0.28	
197	0.7	35.7	− 9.02	− 7.25	− 1.42	− 1.22	− 0.42	Depression of Tikacs
198	0.4	36.1	+61.45	+54.20	6.83	5.61	3.52	(III/3)
199	1.0	37.1	−57.25	− 3.05	− 6.36	− 0.75	− 1.31	
0009	1.9	39.0	+ 0.89	− 2.16	0.10	− 0.65	− 0.01	

TABLE 1. Continued

Number of Point	L (km)	ΣL (km)	ΔH (mm)	ΣΔH (mm)	ΔV (mm/year)	ΣΔV (mm/year)	$\frac{\Delta V}{L}$ ("/km)	Remarks
167	0.3	12.4	- 0.74	- 7.45	- 0.08	- 0.83	- 0.06	Séd stream
168	0.5	12.9	- 5.24	-12.69	- 0.58	- 1.41	- 0.24	(II/3)
169	0.2	13.1	- 6.68	-19.37	- 0.74	- 2.15	- 0.76	
170	0.8	13.9	+12.14	- 7.23	1.35	- 0.80	0.35	Tündér hill
171	0.4	14.3	+23.68	+16.45	2.63	1.83	1.36	(II/4)
172	0.8	15.1	-26.26	- 9.81	-2.92	- 1.09	-0.75	Basin of Berhida
173	1.1	16.2	+ 1.30	- 8.51	0.14	- 0.94	0.03	(II/5)
174	1.2	17.4	- 1.32	- 7.19	- 0.15	- 0.80	- 0.02	
014	0.7	18.1	- 1.07	- 6.12	- 0.16	- 0.68	- 0.04	
175	0.1	18.2	- 1.41	- 7.53		- 0.84	- 0.32	(II/6)
176	1.2	19.4	- 0.65	- 6.88	- 0.07	- 0.76	- 0.01	
177	1.0	20.4	- 3.36	- 3.52	- 0.37	- 0.39	- 0.08	
178	0.7	21.1	- 0.87	- 4.39	- 0.10	- 0.49	- 0.03	
180	1.3	22.4	- 0.22	- 4.61	- 0.02	- 0.51	∅	
015	0.1	22.5	- 2.30	- 2.31	- 0.26	- 0.25	- 0.53	
181	0.5	23.0	- 3.28	- 5.59	- 0.36	- 0.62	- 0.15	(III/1)
182	1.3	24.3	- 0.24	- 5.83	- 0.03	- 0.65	∅	
183	1.0	25.3	- 0.93	- 4.90	- 0.10	- 0.54	- 0.02	
016	1.3	26.6	- 0.23	- 5.13	- 0.03	- 0.57	∅	

TABLE 2. Computed Accuracy of the Changes in Height

L_i (km)	$\mu_{\Delta H_i}$ ±(mm)
0.1	0.14
0.2	0.20
0.3	0.24
0.4	0.28
0.5	0.32
0.6	0.35
0.7	0.37
0.8	0.40
0.9	0.42
1.0	0.45
1.2	0.49
1.5	0.55
1.8	0.60
2.0	0.63
2.5	0.71
3.0	0.77

Tikacs depression:

$$\Delta H = - (9.02 \pm 0.37) \text{ mm}$$
$$\Sigma \Delta H = - (11.25 \pm 2.67) \text{ mm}$$
$$\Delta V = - (1.02 \pm 0.04) \text{ mm/a}$$
$$\text{hor.grad.} = - (0.30 \pm 0.012) \text{ "/a}.$$

- At the section between Berhida and Küngös smaller (and of changing sign) changes in height were obtained, but this area subsided related to Veszprém (in a small scale).

A significant trend of movements was revealed by the investigation from the area of Litér to the Basin of Berhida where clear and significant trends changing in sign correspond to the geological situation.

According to this

- the area of Litér subsided definitely (II/1)

TABLE 3. Standard Errors of Summed Up Changes in Height

$\Sigma_V^i L$ (km)	$\mu_{\Sigma \Delta H}$ ±(mm)
5	1.0
10	1.4
15	1.7
20	2.0
25	2.2
30	2.5
35	2.6
40	2.8

TABLE 4. Standard Errors of Summed Up Changes in Height at Characteristic Places

Denomination		$\sum\limits_{V}^{i} L$ (km)	$\mu_{\Sigma\Delta H}$ ±(mm)
Area of Litér	II/1	10.6	1.46
Mogyorós hill	II/2	12.0	1.55
Séd stream	II/3	12.9	1.60
Tündér hill	II/4	14.1	1.68
Basin of Berhida	II/5	15.6	1.77
Lepsény	–	39.0	2.79

- Mogyorós-hill uplifted related to the basin of Litér (but it subsided related to Veszprém)
- the valley of stream Séd (II/3) subsided significiantly (related both to Veszprém and Mogyorós-hill)
- Tündér-hill (II/4) significantly uplifted related to Veszprém and especially to its surroundings
- Tikacs and the basin of Berhida as living depression are confirmed.

The investigated line intersects the valley of stream Séd at the eastern part of Berhida for a second time. For the bench mark situated there, only one measurement is available, so there was no possibility to determine the change in height.

The data obtained in this investigation and their mean square errors were compared. As $\Sigma\Delta H$, ΔV and hor.grad. are based on the ΔH values, all the statistics showed the same results for the same section. Therefore they show only a comparison of ΔH and corresponding $\mu_{\Delta H}$ values but the conclusions drawn from them are valid ones.

The following values were compiled for each section of the investigated line:

$$\frac{|\Delta H|}{|\mu_{\Delta H}|} = \varepsilon \quad (8)$$

These data were classified and the frequency of the classes determined; the relative frequency was computed as follows:

$$g_i = f_i/n \quad (9)$$

where $n = 51$.

The data of comparison are presented in Table 5. The first column contains the intervals of ε (larger intervals were used for $\varepsilon \geqslant 10$). In the second column the relative frequentcies are shown.

TABLE 5. Statistical Analysis of Changes in Height and Their Standard Error

$\frac{|\Delta H|}{|\mu_{\Delta H}|} = \varepsilon$	f_i	f'_i	$g_i = f_i/n$	$g'_i = f'_i/n'$
0.0 – 1.0	10	0	0.196	0
1.1 – 2.0	9	2	0.176	0.111
2.1 – 3.0	7	4	0.138	0.222
3.1 – 4.0	4	1	0.078	0.056
4.1 – 5.0	0	0	0	0
5.1 – 6.0	1		0.020	
6.1 – 7.0	1		0.020	
7.1 – 8.0	2	2	0.039	0.111
8.1 – 9.0	1		0.020	
9.1 – 10.0	0		0	
10.1 – 20.0	8	4	0.157	0.222
20.1 – 30.0	2	1	0.039	0.056
30.1 – 40.0	2	2	0.039	0.111
40.1 – 100.0	2	2	0.039	0.111
100.1 – 150.0	2	0	0.039	0
	n=51	n'=18	Σ=1.000	Σ=1.000

Such an investigation was carried out not only for the whole investigated line but for the section from the area of Litér to the Basin of Berhida, too. Results corresponting to the latter are shown in Table 5, too.

Data in Table 5 show:

a.) for the whole investigated line:
- for 63 per cent of the levelling sections

$$g_i \geqslant 2.0$$

- for about 17 per cent of the levelling sections:

$$1.0 \leqslant g_i \leqslant 2.0$$

- for 20.0 per cent of the levelling sections:

$$g_i \leqslant 1.0$$

b.) the section between the area of Litér and Berhida shows the following:
- there was no section where

$$g_i \text{ was less the } 1.0$$

- for 11 per cent of the sections:

$$1.0 \leqslant g'_i \leqslant 2.0$$

- for 22 per cent of the sections:

$$2.0 \leqslant g'_i \leqslant 3.0$$

- for 17 per cent of the sections:

$$3.0 \leqslant g'_i \leqslant 10.0$$

- for 22 per cent of the sections:

$$10.0 \leq g'_i \leq 20.0 \text{ and}$$

- for 28 per cent of the sections:

$$20.0 \leq g'_i \leq 150.$$

These facts show that significant trends are seen around the area of Litér and the basin of Berhida being $|\Delta H|$ twice as large as $\mu_{\Delta H}$ at 89 per cent of the sections, and the derived characteristic were three times larger than their standard errors for 67 per cent of the sections.

As the end of this paper, we should like to draw attention to the following: for the two ends of the investigated line the following absolute velocities were got for the second investigation of recent vertical movements of the Carpathian-Balkan Region (Joó et al, 1985)

V_{abs} - Lepsény 0.00 mm/a and
V_{abs} - Veszprém = + 0.3 mm/a

An interpretation of these data shows that based on the CBR results, Lepsény subsided (relative to Veszprém) with a trend of - 0.3 mm/a.

The recent investigation reinforces this trend, but it shows that the value of subsiding is - 6.16 mm/9 year, e.g. $\Delta V = - 0.68$ mm/a.

As a conclusion it can be stated that significant changes in height could be observed as a result of repeated precise levellings. Most significant trends were shown from the area of Litér to Vilonya and at section 1.5 km to east from Vilonya.

References

Ádám L.-Marosi S., (1959): Die physikalische Geographie von Mezőföld. Akadémiai Kiadó, Budapest

Balla Z., (1984): The Carpathian Loop and the Pannonian basin: A kinematic analysis Geophysical Transactions Vol.30. No. 4. 313-353

Horváth F., (1984): Neotectonics of the Pannonian basin and the surrounding mountain belts: Alps, Carpathians and Dinarides. Annales Geophys, 2(2) 147-157.

Joó I. et al., (1979): Map of Recent Vertical Crustal Movements in the Carpatho-Balkan Region; Scale 1:1000000, Budapest

Joó I. et al., (1985): The new Map of Recent Vertical Movements in the Carpatho-Balkan Region, Scale: 1:1000000, Budapest

Joó I.-Czobor Á.-Gazsó M.-Németh Zs. (1986): The Investigation of Recent Crustal Movements in the Carpatho-Balkan Region and the Co-ordination of the works. Scientific Report of the College for Surveying and County Planning, Székesfehérvár, pp. 144.

Joó I., (1987): Die geodätische Untersuchung der vertikalen Einwirkungen des Erdbebens in 1985 in Ungarn im Gebiet von den Siedlungen Berhida-Peremarton am 15. August 1985. (Neubrandenburg, 20-30. April 1987. DDR)

Mahel M., (1974): Tectonics of the Carpathian Balkan Regions, Geological Institute of Dionyz Stur, Bratislava, pp. 453

Pogácsás Gy., (1984): Seismic statigraphic features of the Neogene Sediments in the Pannonian Basin Geophysical Transactions, VOL. 30.

Pospisil L.-Vass D., (1984): Influence of the structure of the Lithosphere of the formation and development of intramontane and back molasse basins of the Carpathian Mountains. Geophysical Transactions, VOL. 30. No. 4. 355-371.

Szeidovitz Gy., (1987): Instrumental observations of aftershocks of earthquake occured on 15. Aug, 1985 at 6h 29min (manuscript) pp. 74.

Ziegler P.A., (1982): Geological Atlas of Western and Central Europe. Shell Internationale Petroleum Maatschappij B.V. The Hague pp. 130.

VERTICAL MOVEMENT OF INDO-GANGETIC PLAINS

C. S. Joshi, A. N. Singh, Manohar Lal

Geodetic & Research Branch, Survey of India,
17 E.C. Road, Dehra Dun—248001 India.

Abstract. Indo-gangetic plains form a very interesting zone between the peninsular shield and the Himalaya in India. While a number of hypothesis have been put forth/discussed regarding the formation of the Himalaya by interaction of Indian Plate, very little has been examined about the state of these in between plains. This paper gives an account of the attempt made to analyse a number of high precision levelling lines, running across these plains and since repeated. Based on a chosen station, analysis has been done and efforts made to plot vertical movement contours. An attempt has also been made to correlate the movement pattern with the known geological and tectonic features. This investigation will throw very useful light on the tectonics of this region.

Introduction

The flat low-lying area of the alluvial plains, approximately 0.777 million sq km, covering the large portion of Sind, Northern Rajasthan, almost whole of Punjab, Uttar Pradesh, Bihar, Bengal and part of Assam, forms the vast Indo-gangetic plains. The plains are bounded by Himalayan mountain ranges in the North and the peninsular shield in the South. The geologists conceive that the Indo-gangetic depression is a downwarp of the Himalayan foreland of variable depth, converted into flat plains by the simple process of alluviation (Wadia, 1973). Underlying the alluvium are the unconsolidated Siwaliks and older Tertiary sediments and below these are more consolidated older formations. The breadth of the plains is about 500 km in the west and about 150 km in the east. The floor of the gangetic plains is corrugated by inequalities and buried ridges. Two such ridges are indicated by geophysical methods - one in the elongation of Aravali axis between Delhi and Haridwar and the other under the Punjab alluvium from Delhi to salt range (Wadia, 1975; Glennie, 1932). Oldham on the other hand postulated the Indo-gangetic trough to be more or less uniform slope dipping northward and rising rapidly at its northern edge towards the Himalaya (Qureshy, 1970). A number of high precision levelling lines had been run across these plains and repeated over the last about hundred years. It was, therefore, considered of great interest to analyse selected levelling lines to bring out vertical movement pattern of these fascinating plains, after separating episodic movements.

Figure 1 gives a simplified map showing the Indo-gangetic plains and the levelling lines selected for the present study. In this map known thrust lines and important physical features have been marked to enable the authors to bring out the division of the region in different blocks considered by them. Epicentres of earthquakes >6.0 magnitude on Richter scale have also been plotted on this map. The authors have chosen Standard Bench Mark at Dhule in the peninsular shield as the origin. All other elevation changes are, therefore, relative to Dhule Standard Bench Mark.

Approach

The simplest way to obtain velocity parameter is to divide the height difference through time span between the two sets of observations. In mathematical terms:

$$V = \frac{Ht_2 - Ht_1}{t_2 - t_1}$$

Here 'Ht_2' is height observed at epoch 't_2', 'Ht_1' is the height observed at epoch 't_1'. This concept, however, assumes a constant rate of velocity and does not take into account the motion caused by accelerated movement or episodic movements caused by some events like earthquakes. The constant rate of velocity does not take into account that the motion of the earth's crust is correlated to tectonic formations in which the area under investigation could be sub-divided by geological considerations. In this paper the concept generated by the principal author in 1984 and since applied by Arur and Singh, 1986a to one levelling line has been further developed by the present authors and applied to this vast area. In order to have a clear picture of the vertical movement pattern of the area under investigation the study has been extended to a geologically stable area in the peninsular shield up to river Tapti. A second degree

Copyright 1989 by
International Union of Geodesy and Geophysics
and American Geophysical Union.

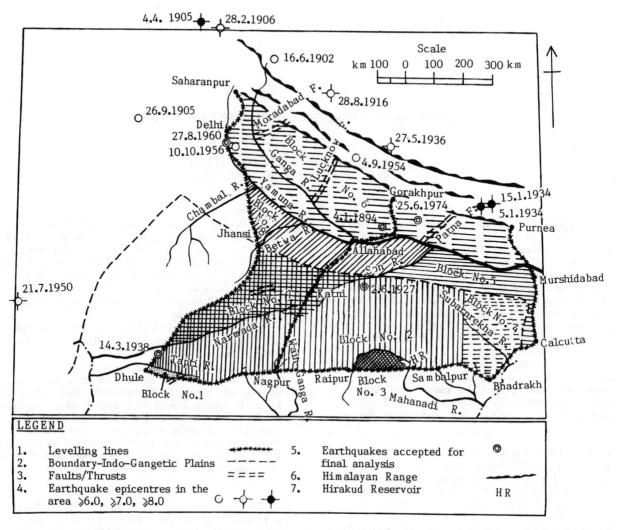

Fig. 1. A simplified map of Indo-gangetic Plains. Legends for the Boundary of Indo-gangetic Plains, levelling lines selected for the study and other details are shown in the map. The region is divided in 8 blocks on geological and physical considerations as shown in the map.

velocity surface has been fitted in the study area making use of velocity at nodal points on all the four levelling profiles indicated in figure 1. The entire area has been divided in 8 blocks on geological and physical considerations. The extent of each block has been indicated in the above figure. Each block is assumed to have its characteristic velocity.

The episodic motions have been separated out and evaluated and movement parameters for each block have been worked out applying the least squares techniques of adjustment. The weights have been attached to the observations vector by calculating these from differences of fore and back levelling observations following the technique suggested by Wassef, [1974] and Arur and Singh, [1986b].

With a view to study the height correlated error and to obtain uniform tectonic tilt associated with each levelling line, profiles of all the four levelling lines have been drawn indicating height changes between two epochs of observations. The topography of the profile has also been drawn for all the four levelling lines. The first derivatives of these two profiles with respect to distance have been worked out in order to evaluate apparent tilt and apparent slope associated with each Bench Mark. Figures 2, 3, 4 and 5 have been prepared for the four levelling profiles separately. On each profile four curves have been drawn. The curve with continuous lines indicates apparent elevation changes vs. distance for the selected Bench Marks (BM). The curve with broken lines gives elevation vs. distance. The third curve gives tilt (derivative

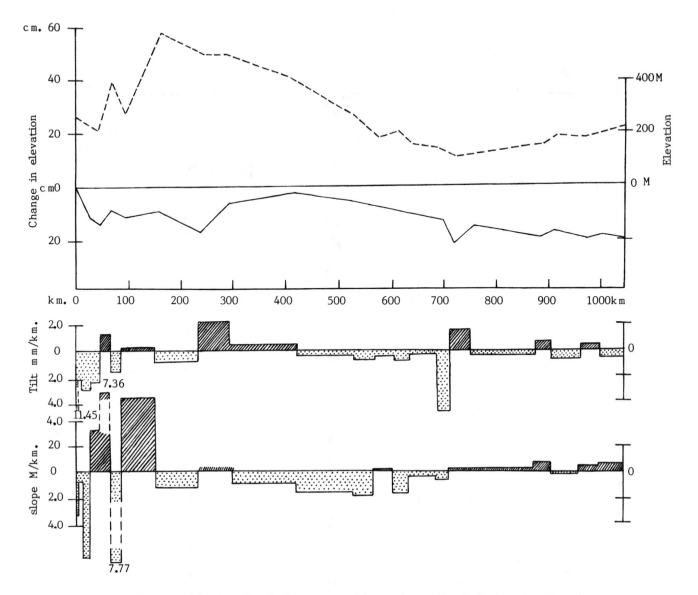

Fig. 2. Levelling data for line No. 1. The curve with continuous line indicates elevation changes vs. distance for the selected Bench Marks (B.Ms.). The curve with broken lines gives relation elevation vs. distance. The third curve gives tilt (derivative of the first curve with respect to distance). The last curve gives the topographic slope (derivative of the second curve with respect to distance).

of first curve with respect to distance). The last curve gives the topographic slope (derivative of second curve with respect to distance), [Jackson et al., 1983].

Mathematical Model

Movement Map

Mathematical Model chosen for the preparation of Movement Map is as under :

$$\Delta H_{a,i} - \Delta H_{a,o} = \Delta t_{i,o}(C_{11}.x_a + C_{21}.y_a + C_{31}.x_a.y_a + C_{41}.x_a^2 + C_{51}.y_a^2) + A_j(x_a, y_a, x_j, y_j, d_j, t_j, t_i, t_o) + r_i$$

Here $\Delta H_{a,i} - \Delta H_{a,o}$ is the change in elevation difference of Bench Mark 'A', run down from starting Bench Mark viz. Dhule, between two epochs of observations viz. t_o and t_i. This is a quasi observable quantity. $C_{11} \ldots C_{51}$ are the velocity coefficients, A_j is the episodic motion coefficient. $C_{11} \ldots C_{51}$ and A_j are the parameters of interest, x_a, y_a are the co-ordinates of Bench Mark 'A' with respect to co-ordinates of initial station (0,0); x_j, y_j are the coordinates of the event (earthquake) and d_j is its focal depth; t_i and t_o are the ith and initial epochs of observations

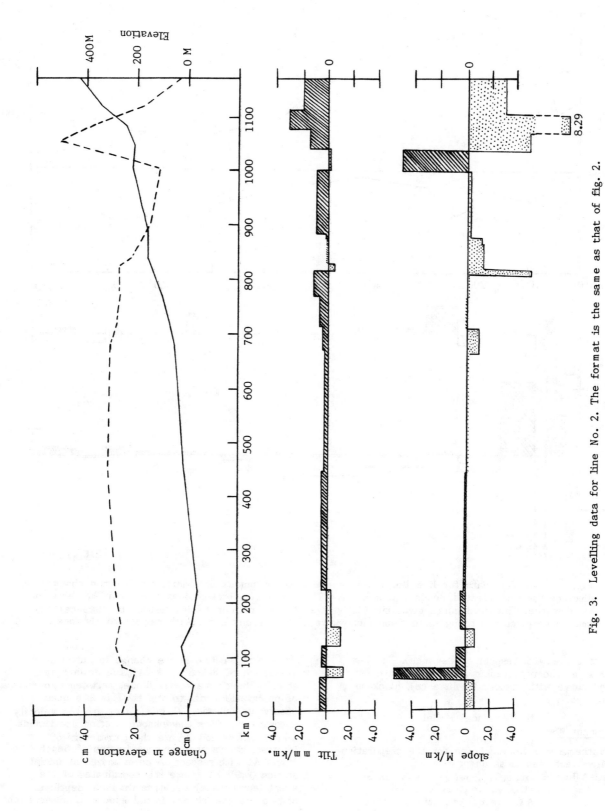

Fig. 3. Levelling data for line No. 2. The format is the same as that of fig. 2.

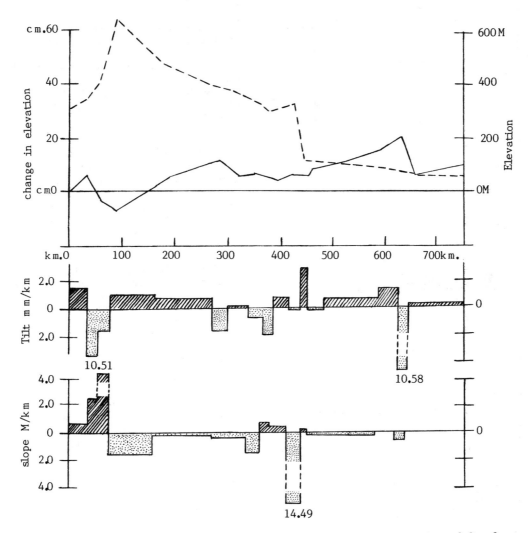

Fig. 4. Levelling data for line No. 3. The format is the same as that of fig. 2.

and ri is the vector of residuals [Arur and Singh, 1986a].

Further Refinements in the Mathematical Model for Episodic Movement Coefficient

In their treatment Arur and Singh, [1980a] had explained the episodic term Aj (xa, ya, xj, yj, dj, tj, ti, to) and other terms of the velocity coefficients. It was brought out that :

Episodic Motion Coefficient
= 0 when ti < tj
= $A_j[(xa-xj)^2+(ya-yj)^2+dj^2]^{-\frac{1}{2}}$
= when ti ≥ tj

The present authors have made further refinements in this model to eliminate all those earthquakes which are at a far-off distance to cause any significant upliftment at the Bench Mark under consideration and accordingly their effect has been considered to be zero. For this purpose the relation $\rho=10^{0.433} \cdot M^{-2.73} \cdot \sqrt[3]{e} \cdot M$ [Mjachkin, et al., 1984] has been used in which 'ρ' is the distance of the Bench Mark from the earthquake epicentre, 'M' is the magnitude of event and 'e' is the deformation expected at the distance 'ρ'. In the levelling observations if we consider the significant deformation level to be 1.4mm being equal to random levelling error propagated in 1 km from the relation $[(1mm/km)^2+(1mm/km)^2]^{\frac{1}{2}}$ for difference of 2 epochs of observations, we can find out the maximum distance of preparation zone of an earthquake to be the distance at which 'e' becomes equal to 1.4×10^6. Considering the elastic property of the earth and keeping in view that the energy released after the earthquake is due to strain built up in preparation zone we can work out the cut off distance after which the effect of earthquake will be zero.

From the above consideration Episodic Motion Coefficient:

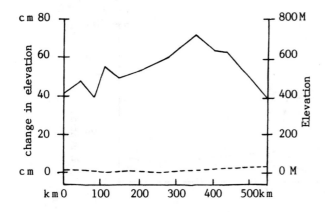
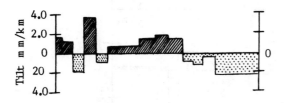

Fig. 5. Levelling data for line No. 4. The format is the same as that of fig. 2.

$$\begin{aligned}
&= 0 \text{ when } t_i < t_j \text{ and } \rho m > \rho j \\
&= A\mathbf{j}[(x_a-x_j)^2+(y_a-y_j)^2+d_j^2)]^{-\frac{1}{2}} \\
&\quad \text{ when } t_i \geq t_j \text{ and } \rho m \leq \rho j \\
\rho m &= 10^{0.433} \cdot M^{-2.73} \cdot \sqrt[3]{e} \cdot M_1 \\
\rho j &= [(x_a-x_j)^2+(y_a-y_j)^2+d_j^2]^{\frac{1}{2}}
\end{aligned}$$

Tilt and slope

In order to study the correlation between the apparent tilt and the apparent slope the mathematical model chosen is as under:

$tli = a + b \cdot si + ei$

Here
- tli = Apparent tilt between the ith pair of Bench Marks.
- si = Geodetic slope between the ith pair of Bech Marks.
- a = Uniform tilt (presumed to be tectonic).
- b = Coefficient of height correlated error.
- ei = Random error.

Data Availability

The availability of repeat levelling data set on all the four lines indicated in figure 1 alongwith year of Survey is given in Table 1. These levelling lines, in various sections, were levelled in different years.

Seven earthquakes with epicentres plotted in figure 1 have occurred since 1860 (the first epoch of observation) in and around the area covered by Indo-Gangetic plains. The earthquake data is given in Table 2.

Analysis of Data

From Table 1 it can be seen that the levelling lines selected for the present study comprise 21 sections observed in different seasons. Considering that error propagation will depend on the terrain through which the levelling profile passes as also on the season in which the levelling has been executed, the authors have worked out actual weights for each levelling line from the analysis of fore and back values of the levelling data for each section as brought out in para 2 i.e. each observation has been assigned weight by calculating these from differences of fore and back levelling observations following the technique suggested by Wassef [1974] and Arur and Singh [1986 b]. The formula used for finding out variances is as under:

$$S_i^2 = \frac{1}{n_i-1} \sum_{j=1}^{n_i} R_{ij}(W_{ij}-W_i)^2$$

Where :

- S_i^2 = Estimate of variance of within the line.
- n_i = Number of sections in line i.
- R_{ij} = Distance of jth section of ith line in km.
- ρ_{ij} = Discrepancy (Fore-Back) of jth section in ith line in metres.

$$W_{ij} = \frac{\rho_{ij}}{R_{ij}}$$

$$W_i = \frac{\sum_{j=1}^{n_i} R_{ij} \cdot W_{ij}}{\sum_{j=1}^{n_i} R_{ij}}$$

All the 16 earthquakes were used for the first analysis of data. 71 permanent bench marks were selected on all the 4 lines keeping in view the No. of blocks and available computer space. We had 56 parameters of interest including 40 parameters for 8 blocks and 16 parameters for the 16 earthquakes. Thus our degree of freedom was fifteen.

After the initial adjustment, however, 7 episodic movement parameters were found to be associated with exclusively high standard errors. Therefore, another adjustment was attempted by excluding the earthquakes associated with these parameters. Subsequent adjustment indicated that three more episodic parameters were showing exclusively high standard errors. Final adjustment was, therefore, done with only five earthquakes. The two earthqakes, one on 10.10.1956 and the other on 27.8.1960 had ocurred in the same locality. These were represented by a single earthquake viz. the

TABLE 1.

Sl. No.	Line No. & its name	Section	Year of old levelling	Year of new levelling
	Line No. 1			
1.	Dhule to Saharanpur	Dhule-Sehore	1883.5	1930.0
2.	"	Sehore-Bhopal	1877.5	1929.0
3.	"	Bhopal-Jhansi	1929.5	1934.5
4.	"	Jhansi-Agra	1925.5	1976.5
5.	"	Agra-Mathura	1905.5	1927.5
6.	"	Mathura-Delhi	1912.5	1928.5
7.	"	Delhi-Meerut	1912.5	1928.5
8.	"	Meerut-Saharanpur	1861.5	1912.5
	Line No. 2			
9.	Dhule to Bhadrakh	Dhule-Nagpur	1933.5	1979.5
10.	"	Nagpur-Raipur	1937.0	1979.5
11.	"	Raipur-Sambalpur	1939.0	1969.5
12.	"	Sambalpur-Bhadrakh	1936.0	1976.5
	Line No. 3			
13.	Nagpur to Gorakhpur	Nagpur-Katni	1908.5	1977.5
14.	"	Katni-Allahabad	1898.5	1977.5
15.	"	Allahabad-Dildarnagar	1864.0	1935.0
16.	"	Dildarnagar-Gorakhpur	1869.5	1935.0
	Line No. 4			
17.	Bhadrakh to Purnea	Bhadrakh-Balasore	1883.0	1937.0
18.	"	Balasore-Howrah	1883.0	1965.0
19.	"	Howrah-Chakdah	1887.0	1965.0
20.	"	Chakdah-Tinpahar	1920.5	1965.0
21.	"	Tinpahar-Purnea	1930.5	1970.0

TABLE 2.

Sl. No.	Date day,month,year	Magnitude	Lat. in degrees	Long. in degrees	Depth in km
1.*	04.01.1894	6.0	26.0	83.0	-
2.	16.06.1902	6.0	31.0	79.0	-
3.	04.04.1905	8.0	32.3	76.2	-
4.	26.09.1905	7.1	29.0	74.0	-
5.	28.02.1906	7.0	32.0	77.0	-
6.	28.08.1916	7.5	30.0	81.0	-
7.*	02.06.1927	6.5	24.0	82.3	-
8.	15.01.1934	8.3	26.6	86.8	-
9.	27.05.1936	7.0	28.4	83.3	-
10.*	14.03.1938	6.3	21.6	75.0	-
11.	29.07.1953	6.0	27.9	82.0	-
12.	04.09.1954	6.7	28.3	83.1	-
13.	21.07.1956	7.0	23.3	70.0	-
14.	10.10.1956	6.7	28.2	77.7	-
15.*	27.08.1960	6.0	28.2	77.4	109.0
16.*	25.06.1974	6.1	26.0	84.3	20.0

*Earthquakes accepted for final analysis.

TABLE 3.

velocity coefficients		standard errors	velocity coefficients		standard errors
Block 1			Block 5		
C11	+0.000003631	±0.000128544	C15	+0.000182873	±0.000225083
C21	-0.000070665	±0.000058319	C25	-0.000374557	±0.001255190
C31	+0.000000240	±0.000004373	C35	+0.000000420	±0.000000738
C41	-0.000000020	±0.000000694	C45	-0.000000140	±0.000000154
C51	+0.000000468	±0.000001745	C55	-0.000000326	±0.000000342
Block 2			Block 6		
C12	+0.000003645	±0.000007175	C16	+0.000003275	±0.000035449
C22	-0.000051174	±0.000013405	C26	-0.000040299	±0.000021793
C32	-0.000000037	±0.000000063	C36	+0.000000012	±0.000000023
C42	+0.000000005	±0.000000006	C46	+0.000000005	±0.000000021
C52	+0.000000219	±0.000000136	C56	+0.000000032	±0.000000016
Block 3			Block 7		
C13	-0.000049746	±0.000725433	C17	-0.000006639	±0.000039174
C23	+0.000586602	±0.014659108	C27	+0.000030248	±0.000070750
C33	-0.000001143	±0.000033113	C37	+0.000000469	±0.000000556
C43	+0.000000073	±0.000001263	C47	-0.000000139	±0.000000166
C53	+0.000003687	±0.000157996	C57	-0.000000456	±0.000000528
Block 4			Block 8		
C14	+0.000069859	±0.000069376	C18	-0.000134788	±0.000087739
C24	-0.001150951	±0.001026155	C28	+0.000115010	±0.000077123
C34	+0.000000954	±0.000000938	C38	-0.000000020	±0.000000068
C44	-0.000000051	±0.000000058	C48	+0.000000136	±0.000000091
C54	-0.000000766	±0.000001177	C58	-0.000000099	±0.000000060

Episodic Movement coefficients associated with 5 earthquakes :

A1 =- 3.5 ± 24.3
A2 =+27.7 ± 39.4
A3 =- 0.9 ± 34.7
A4 =- 7.1 ± 34.7
A5 =- 2.0 ± 0.0

latter one due to lack of repeat levelling data. Results are given in the succeeding paragraph.

Tilt and slope associated with each selected Bench Mark were worked out. Adjusted values of 'a' the regional uniform trend of tilt and 'b' coefficient of height correlated error were worked out. These results are also given in the succeeding paragraph.

Results

The coefficients of regional movement parameters C_{1k}', C_{2k}', C_{3k}', C_{4k}' and C_{5k}' (k indicating block No.) for various blocks and the episodic motion coefficients are given in table 3. Values of tilt for the four lines are given in table 4.

Analysis of Results

Statistical Results

The a-posteriori value of σ_0^2 obtained after the intitial adjustment was nearly 1.3. This value appeared to be quite satisfactory for modern

TABLE 4.

Line No.	Description of line	a=uniform tilt (ppm)*	b=coefficient of height correlated error (ppm)
1.	Dhule - Saharanpur	-0.93±0.57	375±186
2.	Dhule - Bhadrakh	+0.17±0.20	-201±60
3.	Nagpur - Gorakhpur	-0.49±0.73	- 72±167
4.	Bhadrakh - Purnea	+0.17±0.50	-904±3384

*1 ppm = 1 micro-radian.

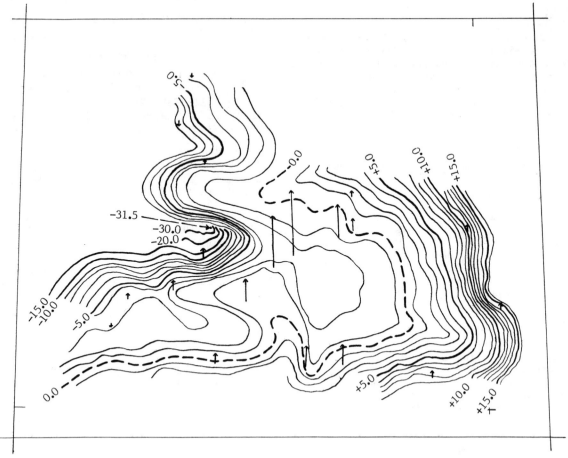

Fig. 6. Vertical Movement contours of Indo-gangetic Plains. Contour interval is 1 mm per year. For episodic motion 1 mm on the map represents 1 cm of vertical movement. The steep contours near Jhansi are shown at contour interval of 10 mm/year and Episodic movements are indicated at a few representative B.Ms.

levelling and indicated good internal consistency of data for both the epochs of observations. The standard errors for the movement parameters associated with Block No. 1 and 3 have been found to be high. These are indicative of inadequacy of data in these blocks. The errors in neglected faults, monument instability, soil motion and errors in seismological data may have also added to the uncertainties in results. More data in space and time domain is bound to improve results.

Vertical Movement Contours

From fig. 6 it is seen that the entire western area is indicating subsidence ranging from 1 mm/year to 31.5 mm/year. The maximum subsidence rate has been found in the area around Jhansi. The negative contours are very steep here. Entire eastern area is indicating upliftment ranging from 1 mm/year to 16 mm/year. Maximum upliftment rate with steep positive contours is seen around Murshidabad. Some steep positive contours are centred around Bhadrakh. An overall picture of NE trend of upliftment can very clearly be seen. It is interesting to note that the Narmada-Son Lineament is associated with minimum rate of velocity with sparse contours.

Episodic Motion

High positive episodic motions of the order of 20 cm are located in the portion Jabalpur to Allahabad from Narmada-Son Lineament to river Ganga. Maximum motion of 20.8 cm is found at Rewah. This portion appears to have released maximum stresses due to earthquakes which have taken place during the intervening period. On the eastern and western flanks positive episodic motions of the order of 3 cm have been found. The B.Ms. in the area enclosed by Mahanadi river are also indicating somewhat high positive episodic movement upto 7 cm. The B.Ms. situated in the portion of Narmada-Son Lineament near Dhule are showing negative episodic movement of the order of 2 cm. Similar trend is seen on the B.Ms. from Mathura to Saharanpur.

Tilt and Slope

Fig. 2 shows that line No. 1 from Dhule to Saharanpur is indicating positive correlation between slope and tilt except at a few places where negative correlation is seen. Fig. 3 shows that line No. 2 from Dhule to Bhadrakh has, at most of places, negative correlation in slope and tilt. Line No. 3 from Nagpur to Gorakhpur is also indicating negative correlation at most of the places as revealed from figure 4. Figure 5 does not show any slope dependent tilt for line No. 4 from Bhadrakh to Purnea since the topography in this area is flat.

Results of Regression Analysis

Although the uniform tilt is not significant at 95% level, in any line, the most significant uniform tilt of −0.93 micro-radians has taken place in the direction Dhule to Saharanpur during an interval of approximately 5 decades. This corroborates with the theory of downwarping of the Indo-Gangetic plains. It appears the downwarping trend is still continuing. A significant height correlated error of +375 ppm appears to have existed in this line. The direction Dhule-Bhadrakh has shown a tilt of 0.17 micro-radians during the same period. However, this tilt does not appear to be significant. A significant height correlated error of −201 ppm appears to have existed in this line. The uniform tilt in the directions Nagpur-Gorakhpur and Bhadrakh-Purnea are −0.49 micro-radians and +0.17 micro-radians respectively. Tilt in these lines also appear to be insignificant. It appears to be extremely improbable that any height correlated errors existed in these two lines.

Conclusions

The present study has revealed subsidence in the Western area and uplift in the Eastern area. Narmada-Son Lineament is associated with minimum rate of velocity with sparse contours. The portion Jabalpur-Allahabad appears to have released maximum stresses due to earthquakes that have taken place during the intervening period. The investigation further revealed that uniform tilts assocated with tectonic processes appear to have taken place along the directions of four levelling lines. The significant tilt in the direction Dhule-Saharanpur suggests that the downwarping of the Indo-gangetic plains is still continuing.

Acknowledgement. The authors are thankful to Shri Atam Prakash of Seismotectonics Cell, Geodetic & Research Branch who has developed the entire computer programme and carefully read the text of the paper. Special thanks are due to Messers K.L. Sawhney and P.K. Uniyal for help in running the computer programme, to Vijay Kumar, V.K. Sharma and Harish Prakash for data reduction and preparation of figures and to N.D. Kothari for typing the text of the paper. The data has been retreaved by the personnel of Seismotectonics Cell and of Levelling Investigation Section of No. 69 Computing Party. The authors are thankful to them.

References

Arur, M.G.; Singh, A.N.; 1986a, Earthquake Prediction from Levelling Data : An Integrated Approach, International Symposium on Neotectonics in South Asia, Dehra Dun, India, Feb. 18-21, 1986, Survey of India, Dehra Dun.

Arur, M.G.; Singh, A.N.; 1986b, Error Analysis of Levelling Data in the Shanan Extension Project in Himachal Pradesh, India, International Symposium on Neotectonics in South Asia, Dehra Dun, India, Feb. 18-21, 1986, Survey of India, Dehra Dun.

Bapat, Arun; Kulkarni, Miss R.C.; Guha, S.K.; 1983, Catalogue of Earthquakes in India and Neighbourhood from Historical period upto 1979, Indian Society of Earthquake Technology, Roorkee.

Glennie, E.A.; 1932, Gravity Anomalies and the Structure of the Earth's Crust, Professional paper No. 27, Survey of India, Dehra Dun.

Gupta, Harsh K.; Nyman, Douglas C.; Landisman, Mark; 1976, Shield-like upper Mantle Velocity Structure below the Indo-gangetic Plains : Inferences drawn from Long-Period Surface Wave Dispersion Studies, Earth and Planetary Science Letters, 34 (1977) 51-55, Elsevier Scientific Publishing Company, Amsterdam-Printed in Netherlands.

Holdahl, S.; 1958, Model for Extracting Vertical Crustal Movement from Levelling Data, Report No. 280, Applications of Geodesy to Geodynamics, Department of Geodetic Science, Ohio State University, U.S.A.

Jackson, David D.; Cheng, ABE; Liu Chi-Ching; 1983, Tectonic Motion and Systematic Errors in Levelling and Trilateration Data for California, Crustal Movements, 1982, Developments in Geotectonics 20, Elsevier, Amsterdam.

Mjachkin, V.L.; Kostrov, B.V.; Sobolev G.A; Shamina O.G.; 1984, The Physics of Rock Failure and its links with Earthquakes, Earthquake Prediction, Proceedings of the International Symbosium on Earthquake Prediction, UNESCO Terra Scientific Publishing Company (TERRAPUB), Tokyo.

Qureshy, M.N.; 1970, Relation of Gravity to Elevation, Geology and Tectonics in India, Proceedings of the Second Symposium on Upper Mantle Project, December, 1970, Hyderabad.

Snay, Richard. A.; Cline, Michal W.; Timmerman, E.L.; 1984, Horizontal Crustal Deformation Models for California from Historical Geodetic Data, International Symposium on Recent Crustal Movements of Pacific Region, Feb. 9-14, 1984, Wellington, New Zealand.

Wadia, D.N.; 1975, "Geology of India", Tata McGrow Hill Publishing Co., New Delhi.

Wassef, A.M.; 1974, On the Search for Reliable Criteria of the Accuracy of Precise Levelling based on Statistical Consideration of the Discrepancies, Bulletin Geodesique No. 112, International Association of Geodesy, Paris.

STRAIN ANALYSIS OF TECTONIC MOVEMENTS FROM GEODETIC DATA ACROSS KROL AND NAHAN THRUSTS

C. S. Joshi, A. N. Singh, Atam Prakash

Geodetic & Research Branch, Survey of India,
17 E.C. Road, Dehra Dun—248001 INDIA.

Abstract. Crustal Movement Studies are continuing in the Kalawar area of Yamuna Hydro-electric Scheme, Dak Pathar, Dehra Dun, India since 1973-74 across Krol and Nahan Thrusts. The data collected during 1973-74, 1975-76, 1978-79 was analysed during 1984 [Arur and Rajal, 1984]. Since then one more set of data has been acquired during 1985-86. In the present paper an attempt has been made to reinvestigate the entire set of data by dividing the region in three principal blocks. The strain pattern has been explained in the context of prevailing tectonic features.

Introduction and Tectonics of the Area

In the Kalawar area of Yamuna Hydro-electric Project in the state of Himachal Pradesh in India, there is a very interesting geological setting around a tunnel of the scheme which passes below this area. Here two important Thrust lines are crossing the tunnel alignment. Fig. 1 gives the Tectonic set-up of the area in which the position of Thrust lines in relation to the position of the triangulation stations in the area has been marked. In between the two Thrust lines are located the Subathu formations which are geologically indicated as Intra-Thrust Zone. The Subathu formations are sandwiched betwen the two Thrusts due to different rates of movement of the Krol Thrust (KT) on the one side and the Nahan Thrust (NT) on the other side. In this area the formations south of NT are mostly sand stones/clay stones, the Subathu formations are mostly shale and the formations North of KT are Mandhalis which are mostly slates/phyllites/quartzites. Horizontal and vertical movement studies are continuing in the area since 1973-74. Observational data for four epochs viz. 1973-74, 1975-76, 1978-79, 1985-86 are available. Data for 1973-74, 1975-76, 1978-79 was analysed during 1984 [Arur and Rajal, 1984].

The analysis of levelling data is not covered under the present study. The authors have applied an entirely new approach suggested by Chrzanowski et al. 1983; Fujii and Nakane, 1983; Welsch, 1983; for working out strain parameters from displacement field and by dividing the region in three principal blocks, one to the SE of the NT, the other to the NW of the KT and the third in the Intra-Thrust Zone. A very clear picture of the strain parameters has emerged from this study.

Fig. 2 gives vectorial presentation of the movement pattern (displacement field) during the three epochs of observations after the initial epoch. The data was used for working out strain parameters.

According to geological interpretation of the Tectonics of the area the KT is older, the NT is younger and the Subathu formations are sandwiched between the two. Geologists view is that KT is moving in the northerly direction at a faster rate than the NT which is also moving in the same direction.

Approach

Stress accumulation in the earth's crust has probably uniform distribution over fairly large areas, at least in the tectonic processes [Caputo, 1978]. It follows that monuments located on one set of geological setting will have similar pattern of strain field. It is, therefore, assumed that the block SE of NT has got one set of translatory motion alongwith a particular pattern of deformation caused by stress field generated in this block. The Subathu formations in the intra-thrust area have translatory movement equivalent to the resultant of the two blocks alongwith another characteristic pattern of deformation caused by stress field generated in this block. Figures 3, 4, 5 and 6 provide sequential representation of strain and rotation parameters alongwith the concept of block dependent movement [Jaeger and Cook, 1969; Chrzanowski, et al., 1983].

Mathematical Model

From the displacement field we have following relations:

$(dx)B = c_0 + (e_{xx})B.x + (e_{xy})B.y - \omega B.y$
$(dy)B = g_0 + (e_{xy})B.x + (e_{yy})B.y + \omega B.x$
$(dx)C = h_0 + (e_{xx})C.x + (e_{xy})C.y - \omega C.y$
$(dy)C = k_0 + (e_{xy})C.x + (e_{yy})C.y + \omega C.x$

Copyright 1989 by
International Union of Geodesy and Geophysics and American Geophysical Union.

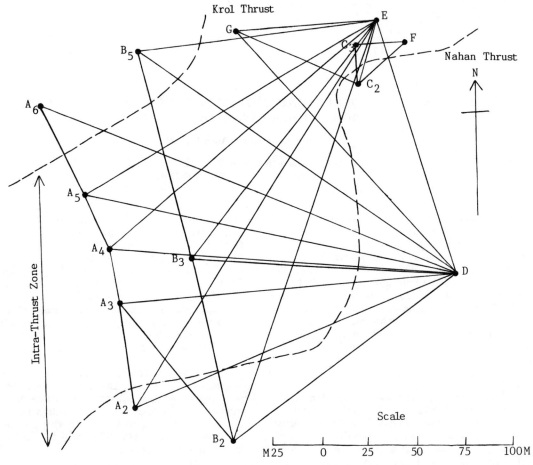

Fig. 1. Tectonic set up of the area and scheme of Triangulation Observations. Block A covers the Intra- Thrust Zone, Block B covers the area SE of Nahan Thrust and Block C covers the area NW of Krol Thrust.

$(dx)A = (co+ho)+(exx)A.x+(exy)A.y-(\omega)A.y$
$(dy)A = (go+ko)+(exy)A.x+(eyy)A.y+(\omega)A.x$
[Chrzanowski et al., 1983; Welsch, 1983; Fujii and Nakane, 1983].

Here $(dx)B$, $(dy)B$ are the displacement field of block B; co, go are translatory motion in block B; $(exx)B$, $(exy)B$, $(eyy)B$ and $(\omega)B$ are the strain and rotation parameters of block B. x and y are the co-ordinates (reduced from origin) of the point under consideration. Similar notations are for other blocks.

Other linear relations from the strain parameters are:

$\gamma_1 = $ eyy - exx   Pure shear
$\gamma_2 = $ 2exy   Engineering shear
$\Delta = $ exx + eyy   Dilatation

This can be written as:

$$\begin{vmatrix}\gamma_1\\ \gamma_2\\ \Delta\end{vmatrix} = \begin{vmatrix}-1 & 0 & 1\\ 0 & 2 & 0\\ 1 & 0 & 1\end{vmatrix} \cdot \begin{vmatrix}exx\\ exy\\ eyy\end{vmatrix}$$

This can be written as:
$P_2 = H_2 \cdot e$

Therefore $\Sigma(P_2) = H_2 \cdot \Sigma(e) \cdot H_2^T$. This will provide variances of elements of (P_2) [Welsch, 1983].

Other non-linear relations from the above linear relations are:

$\gamma = (\gamma_1^2+\gamma_2^2)^{\frac{1}{2}}$ Total shear
$e_1 = \frac{1}{2}(\Delta+\gamma)$ Maximum principal strain
$e_2 = \frac{1}{2}(\Delta-\gamma)$ Minimum principal strain
$\theta = \frac{1}{2}\arctan(\gamma_2/\gamma_1)$ Bearing of e_1
$\psi = \theta+45°$ Bearing of maximum shear

In matrix term by linearization from Taylor's theorem:

$$\begin{vmatrix}\gamma\\ e_1\\ e_2\\ \theta\end{vmatrix} = \begin{vmatrix}\gamma_1/\gamma & \gamma_2/\gamma & 0\\ \gamma_1/2\gamma & \gamma_2/2\gamma & \frac{1}{2}\\ -\gamma_1/2\gamma & -\gamma_2/2\gamma & \frac{1}{2}\\ -\gamma_2/2\gamma^2 & \gamma_1/2\gamma & 0\end{vmatrix} \cdot \begin{vmatrix}\gamma_1\\ \gamma_2\\ \Delta\end{vmatrix}$$

This can be written as:
$P_3 = H_3 \cdot P_2$

Therefore, $\Sigma(P_3) = H_3 \cdot \Sigma(P_2) \cdot H_3^T$. This will provide variances of elements of P_3 [Welsch, 1983].

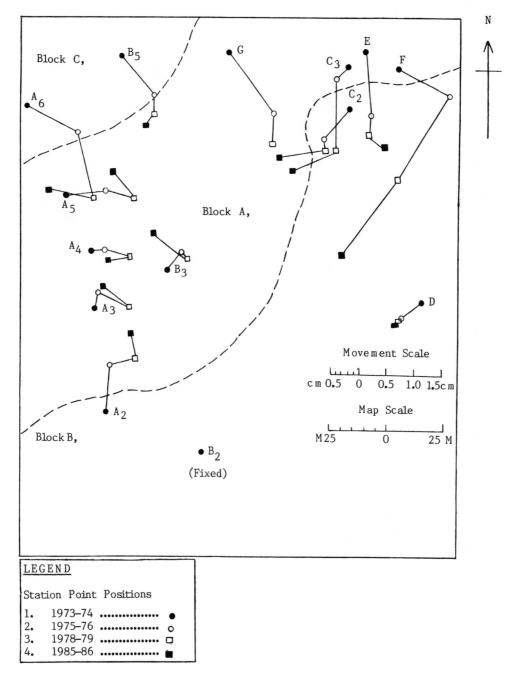

Fig. 2. Vectorial presentation of the movement pattern with B2 held fixed. The station point positions during different epochs are indicated by symbols as shown in the legend box.

Data Availability and its Analysis

As explained under para relating to Introduction and Tectonics of the area, triangulation data in the area is available for four epochs viz. 1973-74, 1975-76, 1978-79 and 1985-86. It was found that two station points viz. B4 and C1 were found destroyed during 1978-79 and accordingly displacement field at these locations were not available during 1978-79 and subsequent epochs. Earlier computations by Arur and Rajal (1984) had used these stations for adjustment of data acquired during 1973-74 and 1975-76. The present authors, however, decided not to consider these

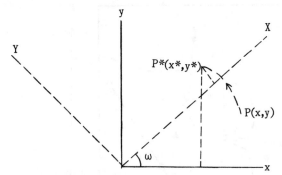

Fig. 3. Rotation of the station point P(x,y) in plan. P is the unrotated position and P* is the rotated position with rotation = ω in anti-clockwise direction.

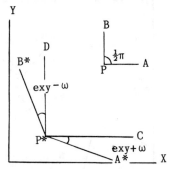

Fig. 5. Representation of Shear Strain. exy is the shear strain parameter. ω is the rotation parameter shown in clockwise direction.

station points for any epoch so that similar constraints are available for adjustment of data during each epoch. Accordingly entire data set was re-adjusted with B_2 fixed for each epoch and displacement field was worked out afresh by taking difference of adjusted co-ordinates obtained during the two epochs of data acquisition. The displacement field thus obtained formed the quasi-observable data-set for subsequent least squares adjustment for estimating most probable value of the parameters viz. elements of strain field. During 1985-86 observations, however, one more monument viz. 'G' was destroyed and accordingly this was omitted for reduction of data during this epoch and displacement at this point could not be worked out for the 1978-79 to 1985-86 epochs.

Results and Interpretation

Statistical Test for Mathematical Model

In order to test the appropriateness of Mathematical Model used for recovery of parameters, the quadratic form of the residuals was employed in a global test, as is usually done in the least squares estimation process. Under this test the null hypothesis is that the variance factor σ^2_{oe} estimated by the adjustment yielding the e, is the same as the a-priori value σ^2_0, from the combination of any pair epochs adjustment for xi, [Chrzanowski et al., 1983].

$H_o : \sigma^2_{oe} = $ versus $HA : \sigma^2_{oe} \neq \sigma^2_0$

The statistics then becomes:

$$T^2 = \frac{\sigma^2_{oe}}{\sigma^2_0}$$

This was compared against the critical value for $F(fe, f, \alpha)$ of the Fisher-distribution, 'fe' being the degree of freedom for estimation of 'e', 'f' being the degree of freedom associated with estimation of σ^2_0 and α is the level viz. 95% in this case. Value of T^2 was worked out and was found to be 0.05, 0.06 and 0.09 for the three epochs of data reduction. The critical value was found to be 2.60. The null hypothesis was accepted since the statistics was well within the critical value.

Results of Strain Field Calculations

Strain parameters were estimated and are given in Table 1.

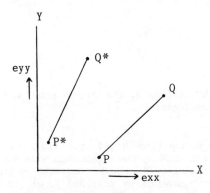

Fig. 4. Strain in length PQ. PQ is the unstrained length and P*Q* is the strained length. exx is the strain component in x-direction and eyy is the strain component in y-direction.

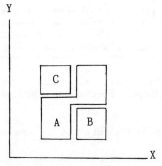

Fig. 6. Representation of block dependent movement. A represents the Intra-Thrust Zone, B the block containing area SE of Nahan Thrust and C the block containing area NW of Krol Thrusts.

TABLE 1. Strain Parameters

Sl. No.	Parameter	1973–74, 1975–76 Estimate	1975–76, 1978–79 Estimate	1978–79, 1985–86 Estimate
Block A : Intra – Thrust Zone				
1.	Strain 'exx' micro-strain/year	– 15±21	– 22±10	– 6±5
2.	Strain 'eyy' micro-strain/year	– 37±49	– 9±17	– 6±9
3.	Shear 'exy' micro-radian/year	4±22	– 3±9	0±5
4.	Rotation 'w' micro-radian/year	– 6±22	– 11±9	– 5±5
5.	Translation E mm/year	– 1.0±2.9	2.7±1.4	–0.6±0.8
	N mm/year	13.0±6.1	1.5±2.1	2.0±1.1
Block B : SE of Nahan Thrust				
1.	Strain 'exx' micro-strain/year	– 8±20	– 15±8	4±4
2.	Strain 'eyy' micro-strain/year	34±19	1±8	4±4
3.	Sheer 'exy' micro-radian/year	– 35±10	1±4	– 8±2
4.	Rotation 'w' micro-radian/year	– 23±10	– 6±4	2±2
5.	Translation E mm/year	– 1.0±0.4	3.2±0.2	0.8±0.1
	N mm/year	9.6±0.4	1.8±0.2	1.4±0.1
Block C : NW of Krol Thrust				
1.	Strain 'exx' micro-strain/year	7±22	– 12±13	14±15
2.	Strain 'eyy' micro-strain/year	9±38	3±13	1±10
3.	Shear 'exy' micro-radian/year	– 8±31	16±14	2±15
4.	Rotation 'w' micro-radian/year	– 24±31	– 5±14	– 14±15
5.	Translation E mm/year	0.0±2.5	–0.5±1.2	– 1.4±0.7
	N mm/year	3.4±5.7	–0.3±1.9	0.6±1.0

Results of Calculation of Other Related Station Parameters

Other related parameters were also worked out and their variances were estimated. The results are given in Table 2.

Analysis of Movement Pattern

Figure 2 represents three deformation patterns worked out at each station point from 4 epochs of data reduction. Each data set has been subjected to adjustment under similar constraints keeping station point 'B$_2$' fixed. It can be seen that station points 'A$_6$', 'B$_5$' and 'G' are having more or less similar pattern of deformations. It is, therefore, quite probable that the KT alignment is passing south of 'G' in place of the geologically indicated alignment vide Fig. 2.

The station points 'B$_2$', 'D' and 'C$_2$' appear to have been well chosen on the NT having similar trends of movement barring some local trends seen at D and C$_2$. Station points 'E', 'F' and 'A' appear to have been constructed on overburden. Most of the station points located on Subathu formations have shown reversal trend in deformation during the last epoch of data analysis as is evident from Fig. 2.

Interpretation of Strain Pattern

The uncertainties of the strain parameters are too high due to scanty data to provide any definite trend in the time domain. Nevertheless, the suitability of the technique for computation of uniform strain parameters for each block is proved beyond doubt. Barring uncertainties of the various

TABLE 2. Other Related Parameters

Sl. No.	Parameter	1973–74, 1975–76 Estimate	1975–76, 1978–79 Estimate	1978–79, 1985–86 Estimate*
Block A : Intra – Thrust Zone				
1.	Pure Shear γ_1	– 23±36	13±16	– 1± 9
2.	Engineering Shear γ_2	7±98	– 6±33	0±18
3.	Dilatation Δ	– 52±23	– 31± 9	– 12± 5
4.	Total Shear γ	24±54	14±23	1±11
5.	Maximum Principal Strain e_1	– 14±24	– 9±13	– 6± 6
6.	Minimum Principal Strain e_2	– 38±34	– 23±11	– 7± 6
7.	Bearing of e_1 clockwise from North	99°±107°	102°±58°	100°±104°
8.	Bearing of maximum Shear clockwise from North	54°±107°	57°±58°	55°±104°
Block B : SE of Nahan Thrust				
1.	Pure Shear γ_1	43±26	16±11	0± 6
2.	Engineering γ_2	– 70±38	2±16	– 15± 9
3.	Dilatation Δ	26±17	– 14± 7	9± 4
4.	Total Shear γ	82±37	16±11	15± 9
5.	Maximum Principal Strain e_1	54±20	1± 4	12± 5
6.	Minimum Principal Strain e_2	– 28±21	– 15± 8	– 3± 4
7.	Bearing of e_1 clockwise from North	119°±10°	87°±30°	134°±11°
8.	Bearing of Maximum shear clockwise from North	74°±10°	42°±30°	89°±11°
Block C : NW of Krol Thrust				
1.	Pure Shear γ_1	2±38	15±19	– 14±20
2.	Engineering Shear γ_2	– 16±76	32±26	4±21
3.	Dilatation Δ	16±38	– 8±19	15±22
4.	Total Shear	16±75	36±26	14±17
5.	Maximum Principal Strain e_1	16±40	14±17	15±15
6.	Minimum Principal Strain e_2	0±44	– 22±15	0±13
7.	Bearing of e_1 clockwise from North	132°±69°	58°±15°	97°±46°
8.	Bearing of maximum Shear clockwise from North	87°±69°	13°±15°	52°±46°

* In units of micro strain/micro radians unless indicated otherwise.

parameters, it is seen that the strain component along E- direction has shown reversal trend in all the blocks whereas strain component along N- direction has shown increasing trend for block A, reversal trend for block B and decreasing trend for block C. Shear strain has shown reversal trend in all the blocks. In this case this trend is more conspicuous. Rotation parameter in case of Subathu formations has shown a regular trend of clockwise rotation thereby indicating it to be composed of uniform material. Rotation parameter of other blocks indicate reversal trend from clockwise rotation to anti-clockwise rotation, thereby indicating that these faults are changing their characteristics from Sinistral to Dextral fault. The pattern may be correlatable with the composition (lithology) of the Subathu formations and that of the Mandhali/Nahan i.e. homogeneous behaviour for the former and heterogeneous for the latter. The translatory motion has also shown a reversal trend in its direction and magnitude in all the blocks. However, NT has shown faster rate of movement than KT which appears to be against the geologically indicated movements.

Interpretation of Other Related Strain Parameters

Since the uncertainties are too high due to scanty data nothing can be said with confidence about the emerging trend. However, the results of these parameters are providing some trend which can be correlated to the lighology of the area. Subathu formations are indicating a regular contraction pattern and the KT and NT are indicating a reversal from expansion to contraction and then again expansion trend. The azimuth from north of maximum shear in Subathu formations has been found to vary from 54° to 57°. The direction of maximum shear in case of KT and NT has, however, shown reversal trend. The direction has changed from 74° to 42° and then to 89° in NT and from 87° to 13° and then to 52° in KT. The regular trends in the case of Subathu formations and the varying trends in case of NT and KT may also be indicative of lithology of these formations.

Correlation with Earthquake

Two earthquakes of intensity 5.5 took place during 1986 one in Dharmsala area in Himachal Pradesh and the other in NW Uttar Pradesh. The latter took place on 16.7.87 (Epicentre : Lat. $31^\circ.25$, Long. $78^\circ.15$) and its effect was felt in Doon Valley. Since the last repeat observation was undertaken after a lapse of about 7 years, it is entirely difficult to assess precisely the time when the reversal trend took place or when the rate of decrease or increase in the value of strain component took place. It is also difficult to say whether or not the reversal pattern or change in rate of change of strain pattern has to do something with the earthquake occurrence in this area since standard errors for various parameters are too high. Seismologists are requested to correlate these data with micro-earthquake data in the area. Repeat observations at frequent intervals and distribution of more points in space would have given a better insight into the changes in strain pattern related to earthquake occurrence.

Comparison of Results

Arur and Rajal [1984] had found the strain from 61 to -185 micro strain in the E- direction and from 370 to -253 in N direction during 1973-74, 1975-76. During the season 1975-76 to 1978-79, these values were reported to range from -98 to 107 and -51.9 to 206.5 respectively. During the present study these values have been worked out for different blocks and definite trend is emerging. The standard errors have been worked out which have, however, been found to be too high. The high value of standard errors are indicative of irregular movements caused by some station points located on overburden which have added to the uncertainties of results. In addition the data points in blocks B and C are extremely scantly. In block C there are only 3 station points and in block B also there are hardly 3 station points in addition to the one held fixed. The present study points to the utility of data analysis by dividing the area in suitable blocks. Distribution of more points in space for each block is bound to improve results and to bring down uncertainties.

Recommendations

Some new station points were selected during 1986. It is recommended that the entire scheme of observations be repeated at more frequent intervals say at least every 2 years, if not every year. In each block a minimum of 5 station points should be located to obtain data redundancy. These station points should be properly anchored to the bed-rock [Arur and Joshi, 1983].

Acknowledgments. The authors are thankful to Dr. M.G. Arur of Survey of India with whom the authors had preliminary discussions. The authors are thankful to Dr. K.N. Srivastava and Dr. S.K. Some of Geological Survey of India for some useful discussions from geological point of view. Special thanks are due to Messers K.L. Sawhney, Vijay Kumar and V.K. Sharma of Seismotectonics Cell, Geodetic & Research Branch for help in running the Computer Programme and reduction of data and to Mr. N.D. Kothari for typing the text of the paper.

References

Ansari, A.R.; Chugh, R.S.; Sinhval H.; Khattri, K.N. and Gaur, V.K.; 1976, Geodetic Determination of Earth Strains and Creep on the Krol Thrust in the Dak Pathar Area, Dehra Dun Distt., U.P. Himalayan Geology Volume 6, Wadia Institute of Himalayan Geology.

Arur, M.G. and Joshi, C.S.; 1983, Suggestions to formulate minimum acceptable observational and computational procedures for RCMS, Proceedings Eighteenth General Assembly of the International Union of Geodesy and Geophysics, August 15-22, 1983, Hamburg, Federal Republic of Germany.

Arur, M.G. and Rajal, B.S.; 1984, Results of analysis of data from Geodetic Surveys for Crustal Movements across Krol and Nahan Thrusts, Proceedings International Symposium on Recent Crustal Movements of the Pacific Region, Feb. 9-14, 1984, Wellington, New Zealand.

Caputo, Michele; 1978, Problems and Advances in Monitoring Horizontal Strain, Application of Geodesy to Geodynamics, Report No. 280, Proceedings of the International Symposium of the Ninth Geodesy/Solid Earth and Ocean Physics (GEOP) Research Conference, October 2-5, 1978, Department of Geodetic Sciences, Ohio State University, Columbus, U.S.A.

Chrzanowski, A.; Chen, Y.Q.; Secord, J.M.; 1983, On the Strain Analyais of Tectonic Movements using Fault Crossing Geodetic Surveys, Developments in Geotectonics 20, Recent Crustal Movements, Elsevier Publications, Amsterdam.

Dey, R.C.; 1980, Some Observations on the Main Boundary Fault in H.P. Himalaya, Himalayan Geology Volume 10, Wadia Institute of Himalayan Geology.

Fujii, Yoichiro; Nakane, Katsumi; 1983, Horizontal Crustal Movements in the Kanto-Tokai District, Japan, as deducted from Geodetic Data,

Developments in Geotectonics 20, Recent Crustal Movements, Elsevier Publications, Amsterdam.

Jaeger, J.C. and Cook, N.G.W.; 1969, Fundamentals of Rock Mechanics, Chapman and Hall Ltd., London.

Kapur, J.N.; Saxena H.C.; 1982, Mathematical Statistics, S. Chand & Co. Ltd., New Delhi.

Savage, J.C.; 1978, Strain Patterns and Strain Accumulation along Plate Margins, Applications of Geodesy to Geodynamics, Report No. 280, Proceedings of the International Symposium of the Ninth Geodesy/Solid Earth and Ocean Physics (GEOP) Research Conference, October 2-5, 1978, Department of Geodetic Sciences, Ohio State University, Columbus, U.S.A.

Welsch, M. Walter; 1983, Finite Element Analysis of Strain Patterns from Geodetic Observations across a Plate Margin, Developments in Geotectonics 20, Recent Crustal Movements, Elsevier Publications, Amsterdam.

VISCOELASTIC DEFORMATIONS AND TEMPORAL VARIATIONS IN THE GEOPOTENTIAL

Roberto Sabadini[1], David A. Yuen[2], Paolo Gasperini[3]

Abstract. We give an overview of the time-dependent changes in the geopotential caused by present-day glacial activities and the current disequilibrium of the Antarctic ice sheet. The time-dependent gravity coefficients are calculated from calculating the transient viscoelastic responses in the mantle for both Maxwell and Burgers' body rheology. We have found that non-axisymmetric contributions to the geopotential are more important for recent glacial retreats than for Pleistocene deglaciation. Sensitivity analyses show that the transient gravitational responses do not vary too much with the assumed rates of melting from present-day glaciers. They depend much more on the parameter values of the Burgers' body rheology, especially the actual ratios between the short- and long-term viscosities.

Introduction

In the last several years the importance of temporal changes in the gravity field through the observation of the orbital motions of LAGEOS I (Yoder et al., 1983; Rubincam, 1984) has been brought forth by the important information gained regarding the viscosity structure of the mantle (Peltier, 1983; Yuen and Sabadini, 1985). In these investigations the source of the forcing has been attributed solely to melting from the last deglaciation. Later it was recognized by Yoder and Ivins (1985) that part of the present-day variations in the Earth's rotation rate, or equivalently J_2 (Yoder et al., 1983), may be produced by the present-day retreat of temperate latitude glaciers (Meier, 1984). This point was quantified by Gasperini, Sabadini and Yuen (1986), who also emphasized the potential importance of the Antarctic ice sheet on inducing discernible changes of the sign as the observed J_2 signal. Recent investigations (Yuen et al., 1987) also find that present-day glacial activities and the current variability of the Antarctic ice volume can cause variations in the long-wavelength components of the gravity field as a consequence of transient viscoelastic responses in the mantle. Effects on higher degree harmonic perturbations of the gravitational field from present-day glacial forcings have been investigated by Sabadini, Yuen and Gasperini (1988).

[1]Dipartimento di Fisica, Univ. di Bologna, Italy.
[2]Minnesota Supercomputer Institute, and Dept. of Geology and Geophysics, Univ. of Minnesota, Minneapolis, MN 55415.
[3]Istituto Nazionale di Geofisica, 00161 Rome, Italy.

Copyright 1989 by
International Union of Geodesy and Geophysics
and American Geophysical Union.

The assumption of steady-state rheology may not be correct for timescales of $0(10^2)$ yr, as in recent glacial retreats. The issue of transient rheology has been revived by the discrepancy in the lower mantle viscosity estimates between postglacial rebound signatures (Cathles, 1975; Peltier and Andrews, 1976) and long-wavelength geoid anomalies (Hager et al., 1985). Sabadini, Yuen and Gasperini (1985) were the first to call attention to the problem of the contamination of the steady-state viscosity form transient rheological effects of using a Burgers' body rheology. Peltier (1985) also came to these same conclusions. All of these recent developments have indeed affected our thinking about the interaction between short- and long-term mantle rheology and other means of extracting mantle viscosity than the traditional method of analysis of postglacial rebound signatures. In this paper we give an account of the current status of temporal changes of the gravity field from present-day cryospheric forcings and present some of the sensitivity analysis of the time-dependent gravity coefficients, which were not discussed in our previous works.

Rheological Models

Rheological models used here are linear viscoelastic in nature because we have used the analytical development given by Yuen et al. (1982; 1986) and Sabadini et al. (1982). Another reason for using a linear rheology is because of its ability to fit a wide range of geophysical data. Moreover, our theoretical formulation can handle all types of linear viscoelastic rheologies, ranging from the Maxwell (Sabadini et al., 1982) to the Burgers' body rheology (Yuen et al., 1986).

We will focus especially on the Burgers' body rheology because of its versatility in treating the dynamics of both intermediate and long term geological phenomena. This linear rheology is simple in that it can be completely described by the short- and long-term viscosities v_2 and v_1 and by the relaxed and unrelaxed shear moduli μ_2 and μ_1 (Yuen and Peltier, 1982). By means of the correspondence principle we can put in the shear modulus all of the physical attributes present in transient and steady-state creep. In the Laplace transformed domain the rigidity $\mu(s)$ is given by

$$\mu(s) = \frac{\mu_1 s\left(s + \frac{\mu_2}{v_2}\right)}{s^2 + \left[\frac{(\mu_1 + \mu_2)}{v_2} + \frac{\mu_2}{v_1}\right]s + \frac{\mu_1 \mu_2}{v_1 v_2}} \quad (1)$$

Transient rheology data interpreted in terms of Burgers' body rheology were given by Smith and Carpenter (1987). The shear modulus for the standard linear solid (S.L.S.) can be obtained from (1) by taking the limit v_1 going to infinity. It becomes

$$\mu(s) = \frac{\mu_1\left(s + \frac{\mu_2}{v_2}\right)}{s + \left[\frac{\mu_1 + \mu_2}{v_2}\right]} \quad (2)$$

On the other hand, the modulus for the Maxwell solid commonly employed in modelling of glacial isostasy can be obtained from (1) in the limit s tending toward zero

$$\mu(s) = \frac{\mu_1 s}{s + \frac{\mu_1}{v_1}} \quad (3)$$

In this work we will employ linear rheologies, given by eqns. (1) to (3) and will not consider any rheologies with a continuous spectrum as in the frequency-dependent $\mu(s)$ (Sabadini, Yuen and Widmer, 1985) which can be used to study anelastic phenomena.

We use a four-layer model consisting of an elastic lithosphere, a two-layer viscoelastic mantle consisting of the rheologies given above, and an inviscid core. The boundary between the upper and lower mantles is situated at the 670 km seismic discontinuity. We give the physical parameters in Table 1.

Recent Glacial Forcings

Today we are all familiar with the nature and extent of the melting of the great continental ice sheets which last reached their maximum volumes about 18,000 years ago. The dynamical effects induced by this surface loading event left many discernible geophysical signatures which have been employed to understand mantle viscosity (e.g. Wu and Peltier, 1983). Less known and appreciated, however, is the amount of ongoing glacial forcing from discharges due to valley glaciers (Meier, 1984). Also of tremendous importance is the question involving the mass balance of the Antarctic ice sheet (D.O.E. Report, 1985), which have climatological implications. We have modelled the thirty-one glaciers tabulated in Meier (1984) as point-source forcings, in which the geographical locations are supplied as part of the input function (Gasperini et al., 1986). In the case of Antarctica we have also treated it for excitation of J_2 as a point-source. For higher harmonics we have employed a finite spherical ice-cap with an angular amplitude of twenty degrees (Sabadini et al., 1988). To monitor variations in sealevels at sites along continental margins would require more detailed modelling techniques, such as spectral (Nakada and Lambeck, 1988) or finite-element (Peltier, 1988) methods.

Time-dependent Gravitational Coefficients

In this section we will develop some of the formulas which have been used (Yuen et al., 1987; Sabadini et al.; 1988) to calculate time-varying gravitational coefficients from both point-source and finite-sized disks. The calculations of the transient viscoelastic responses, due to glacial forcings, follow exactly the same procedures as for Pleistocene deglaciation set down in previous papers (e.g. Yuen et al., 1982; Sabadini et al., 1982).

Depending on the timescale under study, the geopotential can be separated into a static background contribution and a time-dependent portion, very small in magnitude compared to the background potential, which may be caused by internal or external forcings. This aspect is the focus of this paper. We can write down the time-dependent perturbation of the geopotential from viscoelastic creep in the mantle as the real function

$$\delta U(r, \theta, \phi, t) = \text{Re}\left[\frac{GM}{r}\sum_{n=2}^{\infty}\sum_{m=-n}^{n}\left(\frac{d}{r}\right)^n J_{nm}(t) Y_{nm}(\theta, \phi)\right] \quad (4)$$

where G is the gravitational constant, M the mass of the Earth, d is the radius, θ and ϕ are respectively the co-latitude and -longitude, t is the time, J_{nm}'s are the gravitational coefficients and Y_{nm}'s are the unnormalized spherical harmonics (Lambeck, 1980). The complex coefficients $J_{nm}(t)$ depend on the forcing function and the eigenspectral properties of the viscoelastic model. In terms of the perturbations to the Stokes coefficients, eqn. (4) may be rewritten as

$$\delta U(r, \theta, \phi, t) = \frac{GM}{r}\sum_{n=2}^{\infty}\sum_{m=0}^{n}\left(\frac{d}{r}\right)^n (\delta C_{nm}(t)\cos m\phi + \delta S_{nm}(t)\sin m\phi) P_{nm}(\cos\theta) \quad (5)$$

TABLE 1. Physical Properties of the 4-layer Earth Model

Physical Property	Value
density of lithosphere and upper mantle	4,120 kg/m³
density of lower mantle	4,580 kg/m³
density of core	10,925 kg/m³
surface gravity	9.7 m/s²
lithospheric rigidity*	7.28 x 10¹⁰ N/m²
upper mantle rigidity*	9.54 x 10¹⁰ N/m²
lower mantle rigidity*	1.99 x 10¹¹ N/m²

*instantaneous rigidity as applicable for seismic waves

where δC_{nm} and δS_{nm} are the perturbations to the Stokes coefficients and $P_{nm}(\cos\theta)$ is the unnormalized associated Legendre function. For a configuration of N point-source loads, the time-dependent Stokes coefficients as a consequence of viscoelastic creep in the mantle are given by

$$\delta C_{nm}(t) = \sum_{i=1}^{N} \frac{2M_i}{M} \int dt'\, f_i(t')(1+k_n(t-t')) \times \frac{(n-m)!}{(n+m)!} P_n^m(\cos\theta_i)\cos m\phi_i \quad (6a)$$

$$\delta S_{nm}(t) = \sum_{i=1}^{N} \frac{2M_i}{M} \int dt'\, f_i(t')(1+k_n(t-t')) \times \frac{(n-m)!}{(n+m)!} P_n^m(\cos\theta_i)\sin m\phi_i \quad (6b)$$

For the zonal-harmonic ($m = 0$) coefficient δJ_n's, the expression takes the form

$$\delta J_n(t) = \sum_{i=1}^{N} \frac{M_i}{M} \int dt' (1+k_n(t-t')) f_i(t') P_n(\cos\theta_i) \quad (6c)$$

The initial mass of the individual glacier is given by M_i, k_n is the time-dependent loading Love number of degree n and $F_i(t)$ is the forcing history of the individual glacier. The individual location of the point-source glacial forcings is given by the geographical coordinates θ_i and ϕ_i.

Next we develop the time-dependent geopotential coefficients from surface loading of a disk with finite angular amplitude α. We begin by writing the surface density of a finite disk with mass M_i (Farrell, 1972)

$$\sigma(\gamma) = \frac{M_i}{4\pi d^2} \sum_{n=2}^{\infty} \frac{(2n+1)(1+\cos\alpha)}{n(n+1)} \frac{\partial P_n(\cos\alpha)}{\partial \cos\alpha} P_n(\cos\gamma) \quad (7)$$

where γ is the angular distance between the observation point and the center of the ice cap. We now introduce the Green's function for the geoid perturbation for a point-source, impulsive load on the surface (Sabadini et al., 1988)

$$J(\gamma, t) = \frac{1}{M} \sum_{n=2}^{\infty} (1 + k_n^L(d,t)) P_n(\cos\gamma) \quad (8)$$

From this Green's function we may obtain the change in the geopotential due to surface loading from all of the N disks as

$$\delta J(\theta,\phi,d,t) = \sum_{i=1}^{N} \int dt' \int d\Omega'\, \sigma_i(\theta'-\theta_i, \phi'-\phi_i) \times f_i(t') J(\theta-\theta', \phi-\phi', t-t') \quad (9)$$

Substituting in the surface densities for N disks into eqn. (9), we obtain for the perturbation to the gravity field

$$\delta J(\theta,\phi,d,t) = \sum_{i=1}^{N} M_i \sum_{n=2}^{\infty} \frac{(1+\cos\alpha_i)}{n(n+1)} \times \frac{\partial P_n(\cos\alpha_i)}{\partial\cos\alpha_i} \int dt'\, f_i(t') J_n(t-t') \times \sum_{m=-n}^{n} \frac{(n-m)!}{(n+m)!} Y_{nm}(\theta_i,\phi_i) Y_{nm}(\theta,\phi) \quad (10)$$

where J_n is given by

$$J_n(t) = \frac{1}{M}\left(1 + k_n^o + \sum_{j=1}^{M} k_n^j e^{s_j t}\right) \quad (11)$$

The residues of the loading Love numbers are given by k_n^o (elastic) and k_n^j (viscoelastic). The inverse relaxation times are given by $\{s_j\}$. For the four-layer Burgers' body models with internal buoyancy present at 670 km depth, the number of modes is given by M = 13 (Sabadini et al., 1988).

From eqn. (10) we can write the time-dependent Stokes' coefficients for N finite disks including the surrounding oceans for a closed hydrological cycle as

$$\delta C_{nm}(t) = \sum_{i=1}^{N} \frac{2M_i}{n(n+1)} \frac{\partial P_n(\cos\alpha_i)}{\partial\cos\alpha_i} \int dt'\, f_i(t') J_n(t-t') \times \frac{(n-m)!}{(n+m)!} P_n^m(\cos\theta_i)\cos m\phi_i \quad (12a)$$

$$\delta S_{nm}(t) = \sum_{i=1}^{N} \frac{2M_i}{n(n+1)} \frac{\partial P_n(\cos\alpha_i)}{\partial\cos\alpha_i} \int dt'\, f_i(t') J_n(t-t') \times \frac{(n-m)!}{(n+m)!} P_n^m(\cos\theta_i)\sin m\phi_i \quad (12b)$$

Analytical functions for f(t) were used by Gasperini et al. (1986) and Sabadini et al. (1988) to describe recent glacial forcings. They take the form

$$f(t) = H(t) - \frac{(t-b)}{a} H(t-b) \quad (13)$$

where H(x) is the Heaviside function, "a" is the duration of recent glacial activities and "b" is the time period in which there is a hiatus of glacial melting. Values of a = 200 yr and b = 500 yr have been employed in Gasperini et al. (1986) and Sabadini et al. (1988). It turns out that variations in b would not change results too much, for b greater than 100 yr (Sabadini et al., 1988).

118 VISCOELASTIC DEFORMATIONS

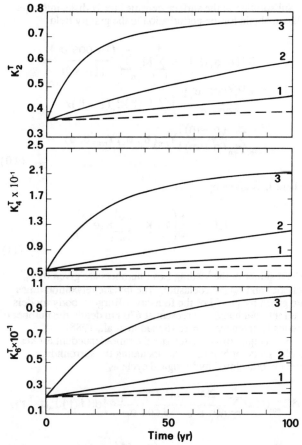

Fig. 1. Time dependence of tidal Love number for degree 2, 4 and 6. Solid curves represent Burgers' body rheology, while dashed curves denote Maxwell rheology. Long-term viscosity of upper mantle is 10^{22} P. Lower mantle steady viscosity is fixed to 3×10^{22} P. Viscosity ratio $v_2/v_1 = 0.01$ for curve 3, 0.1 for curve 2 and 0.5 for curve 1. Rigidity ratio μ_2/μ_1 is 0.1.

Results

We begin this section on results by going over a geophysical signature closely connected with J_2. This is the dispersion of the degree two tidal Love number $k_2^T(t)$, which, with the continual acquisition of satellite data, will better be determined. We will also study the time-dependent tidal Love numbers for higher harmonics and investigate the effects of varying the short-term viscosity.

From analyzing long-period tidal perturbations on the LAGEOS orbital track (Lambeck and Nakiboglu, 1983), it has been proposed that there are some dispersive effects of k_2^T. The accuracy of the observed relation at 18.6 years in the k_2 is probably less than 25% (Merriam, 1985). Their investigation with a frequency-dependent absorption band rheology (Anderson and Minster, 1979) indicated that there is about 20% of anelastic dispersion at a period of 18.6 yr. However, the long-term cut off time in the retardation spectrum suggested by this finding, would produce seismic Q an order of magnitude greater than the observation for the toroidal modes (Sabadini et al., 1985). There remain indeed many outstanding questions in regard to the short-term rheological behavior of the mantle for timescales between 1 and 10^2 years. In Fig. 1 we display variations of the tidal Love numbers for harmonic n = 2, 4, ad 6 and for both the Maxwell and Burgers' body rheologies. A static tidal load has been imposed in this sensitivity study. The rigidity ratio μ_2/μ_1 in Burgers' body rheology is fixed to 0.1, while three different ratios of v_2/v_1 have been used. Solid curves 1,2,3 denote values of v_2/v_1 from 0.5 to 0.01, while the dashed curve represents a Maxwell rheology with upper mantle viscosity of 10^{22} P and a lower mantle viscosity of 3×10^{22} P.

We observe that during the first 100 years elastic behavior still dominates in the case of the Maxwell model (dashed curve), while greater viscoelastic dispersions are exhibited by the Burgers' body curves. One finds that for the Burgers' body rheology smaller values of v_2/v_1 produce the greatest

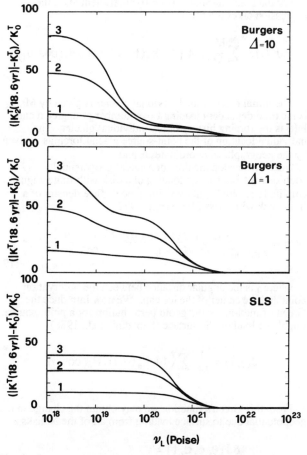

Fig. 2. Percentage of viscoelastic dispersion of tidal Love number with 18.6 yr period as a function of short-term viscosity v_2. Δ is defined by μ_1/μ_2. Curves (1) stand for n = 2, curves (2) for n = 4 and curves (3) for n = 6. Viscosity ratio $v_2/v_1 = 0.1$ is used. S.L.S. rheology is used for bottom panel, while Burgers' body is used for the top two.

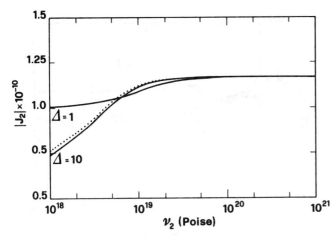

Fig. 3. Changes in J_2 from annual snow load. Solid curve corresponds to Burgers' rheology while dotted curve refers to standard linear solid (S.L.S.). Δ denotes μ_1/μ_2. Viscosity ratio ν_2/ν_1 is fixed to 0.1.

$J_2 = -0.35 \times 10^{-10}$ yr^{-1} has been derived by Yoder et al. (1983) from the LAGEOS data. This particular datum has been used to constrain the viscosity of the lower mantle (Peltier, 1983; Yuen and Sabadini, 1985). Inspection of the figure shows that there exists the possibility for multiple ν_{LM} solutions for the same J_n datum. It is well known that the rotational data such as J_2 and polar motion (O'Connell, 1971; Dickman, 1977) admit more than one solution for the mantle viscosity. Figs. 4 and 5 show that multiple solutions also exist for higher harmonics. Curves with Maxwell rheology (dashed) are very similar to the Burgers' body model with $\nu_2/\nu_1 = 0.1$ and μ_2/μ_1 os decreased to 0.1 (solid curves). Solutions for m = 1 are diminished in amplitude and also exhibit multiple solutions. What is clear is that with the introduction of the Burgers'

amounts of dispersion. These curves in Fig. 1 show definitely that there are pronounced differences in the prediction between Maxwell and Burgers' body rheologies, based on experimentally extracted parameter values (Smith and Carpenter, 1987), for timescales less than a century. We must await for future satellite measurements of $k_n^T(t)$ for the low degree harmonics to constrain further parameter values of short-term mantle rheology. We explore this aspect further by computing the percentage of dispersion defined here to be $(|k_2^T(T = 18.6 \text{ yr})| - k_2^o)/k_2^o$, where k_2^o is the degree-two elastic Love number and T represents the period of the periodic tidal forcing. Fig. 2 shows the amount of dispersion as a function of the short-term viscosity ν_2 for both the Burgers' body (top two panels) and the S.L.S. (bottom panel) models. For large values of Δ ($\Delta = \mu_1/\mu_2$) the abrupt transition from large dispersion to elastic behavior occurs at lower values of ν_2, around 0 (10^{19} P), than in the case for $\Delta = 1$ (middle panel) or for the S.L.S., where the transitional value for ν_2 is about 10^{21} Pa s (Sabadini et al., 1985b).

The effects of snowload on the earth's rotation and gravity fields have recently been examined by Chao et al., (1987) within the context of an elastic model. We show in Fig. 3 that the elastic rheology is essentially valid for ν_2 greater than this cited value. However, in the presence of a pronounced asthenosphere, deformations with shorter wavelengths may be affected to a greater extent than what has been shown for J_2 excitation from annual snow loading.

In Fig. 4 we examine the effects of different rheologies and rheological parameters on the perturbations to the first few even harmonic coefficients of the geopotential from recent glacial activities and the Pleistocene deglaciation, in which the forcing functions from both Laurentide and Fennoscandia are taken into account (e.g. Sabadini et al., 1982). Fig. 4 shows the J_n predictions for a suite of viscosity models in which the lower mantle viscosity is varied between 10^{21} and 10^{25} P. The time of observation is taken to be 80 years following the onset of glacial retreats, which are assumed to have started at 1900 A.D. (Meier, 1984; Gasperini et al., 1986). A value of

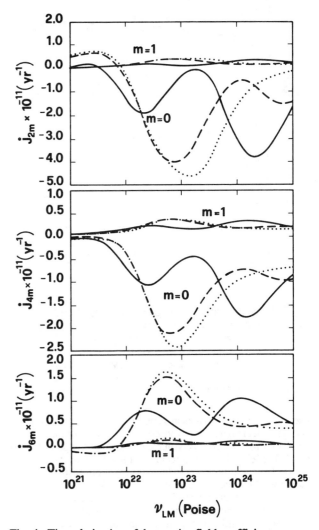

Fig. 4. Time-derivative of the gravity-field coefficient as a function of the lower-mantle viscosity. Time is taken 80 years into the recent glacial activities in the 20th century. Solid curves denote Burgers' body with $\nu_2/\nu_1 = \mu_2/\mu_1 = 0.1$; dotted curves represent $\nu_2/\nu_1 = 0.1$, $\mu_2/\mu_1 = 1$ and dashed curves are derived from Maxwell's rheology. Forcings are derived from recent glaciers and Pleistocene deglaciation.

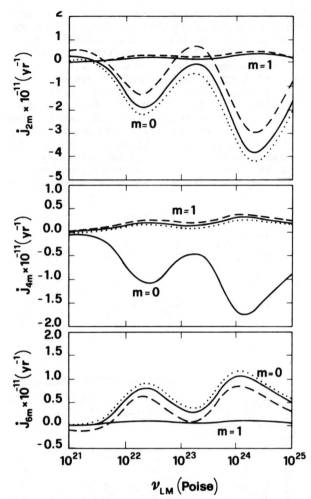

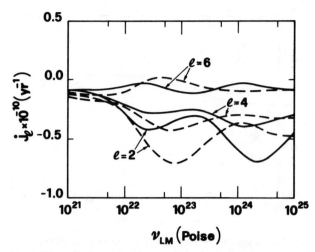

Fig. 6. Time-derivatives of zonal harmonic coefficients as a function of the lower-mantle viscosity. Forcings are due to Pleistocene deglaciation and the growth of Antarctic ice sheet with an equivalent sea-level rise rate of -0.6 mm/yr. Dashed curves represent Maxwell rheology; solid curves denote Burgers' body $\nu_2/\nu_1 = \mu_2/\mu_1 = 0.1$.

Fig. 5. Time-derivative of the geopotential coefficient as a function of the lower-mantle viscosity. Same time window is taken as in Fig. 4. ν_2/ν_1 and μ_2/μ_1 are equal to 0.1. Solid curves have melting rates corresponding to 0.46 mm/yr rise in sea-level; dashed curves have 0.23 mm/yr and dotted curves have 0.92 mm/yr. Forcings are due to current glacial activities and Pleistocene deglaciation.

body, greater complexities are present in the interpretation of geodynamical signatures by viscosity parameterization. Data on glacier balance and ice volume changes for the period 1900 to 1961 (Meier, 1984) results in a sea level rise of 0.46 mm/yr. We will now test the sensitivity of our results to variation of the melting rate by halving it (dotted curve) and doubling it (dotted curve). In Fig. 5 these curves are compared with the predictions from using the nominal rate of 0.46 mm/yr (solid curves). Inspection of Fig. 5 shows that J_n signatures for this time window are not too sensitive to variations of the melting rate.

Studies of the current mass balance of the large Antarctic ice sheet (DOE, 1985) suggest that it may be growing and taking water from the oceans. As the Antarctic ice sheet lies close to the rotational axis, any significant mass exchange between the continent and ocean would have a strong effect on the gravity

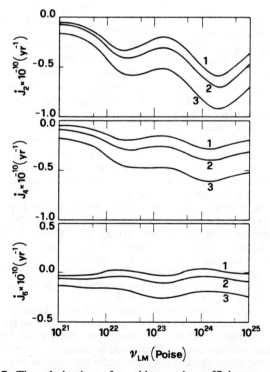

Fig. 7. Time-derivatives of zonal harmonic coefficients as a function of the lower mantle viscosity. Forcings are the same as in Fig. 6, except for varying rates of growth assumed for the Antarctic ice sheet. Curves (1) stand for -0.3 mm/yr; curves (2) for -0.6 mm/yr and curves (3) for -1.2 mm/yr. The rheology is Burgers' body with $\nu_2/\nu_1 = \mu_2/\mu_1 = 0.1$.

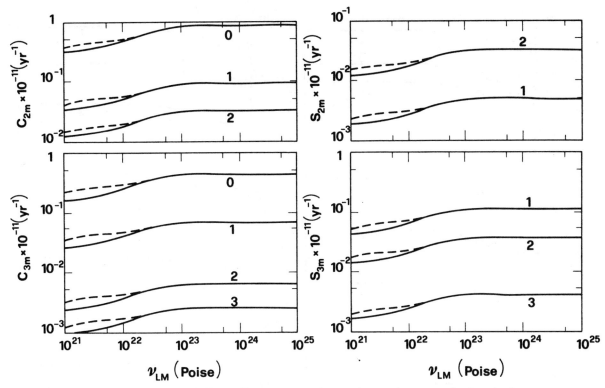

Fig. 8. Secular variation of the Stokes' coefficients with the steady-state lower-mantle viscosity from present-day glacial forcings. Solid curves denote Burgers' body with $v_2/v_1 = \mu_2/\mu_1 = 0.1$, while dashed ones represent $v_2/v_1 = 0.1$ and $\mu_2/\mu_1 = 0.5$. The time is 80 yrs into the period of glacial melting. The steady-state upper mantle viscosity is 10^{22} P. Degrees n = 2,3 are considered.

coefficients $J_n(t)$ with long wavelengths. We have modelled the Antarctic continent as a finite disk with angular radius $\alpha = 20°$. Eqns. (10) and (12) have been used in the calculations of the changes in the gravity field. We examine the responses from the combined forcings of Antarctica growing today and the past Pleistocene deglaciation. In Fig. 6 are displayed the J_n signatures produced by Antarctic ice sheet growing at a rate of - 0.3 mm/yr in the global sea level (DOE, 1985). The time history of this growth is that this phase began in 1900 A.D. and since then has been growing at this same rate. The growth of Antarctica excites both J_2 and J_4 with the same negative sign as for the Pleistocene melting (Yoder et al., 1983). This reinforces the J_2 and J_4 excitation. For this time window there are large differences between the Maxwell (dashed curves) and the Burgers' body rheology (solid curves).

As the present growth rate of the Antarctic ice sheet is still relatively unknown, we have studied the influences of different rates on the J_n signatures. The results for J_2, J_4 and J_6 are shown in Fig. 7 where rates of -.3 mm yr^{-1} (curve 1), -.6 mm yr^{-1} (curve 2) and -1.2 mm yr^{-1} (curve 3). The gravity signals increase with faster rates of growth, although in a weakly nonlinear fashion.

Finally we turn our attention to the excitation of the Stokes coefficients (eqn. (6a) and (6b)) from recent glacial forcings. Figs. 8 and 9 show the predicted geopotential harmonic variations as a function of the steady-state, lower-mantle viscosity. The time is taken again to be 80 yr into the actual glacial retreat period. Both Maxwell (dashed curves) and Burgers' body (solid curves) rheologies have been employed. For m = 2 the zonal harmonic C_{20} dominates over the other tesseral harmonics. For n = 3, zonal harmonic S_{30} again is the leading term. However, for n = 4, non-axisymmetric component (m = 1) becomes the dominant term. Even the order m = 2 term S_{42} is comparable in magnitude to the zonal contribution S_{40}. Extending the calculations to higher degree harmonics, up to n = 6 we find that S_{51} is bigger than C_{50} and that S_{61} is comparable to C_{60}. In general, these results for n up to 6 show that non-zonal gravity coefficients can be comparable or greater than the zonal coefficients. This aspect may allow for some constraints to be imposed on the amount of lateral variations in mantle viscosity. These results also reveal that similar variations of the gravity coefficients with the lower mantle viscosity are exhibited by the higher degree harmonics.

Concluding Remarks

On the basis of the results of these calculations, we argue for the need to monitor current cryospheric activities with modern methods of space observation to address questions in mantle rheology that previously have relied solely on signatures produced by the Pleistocene deglaciation or by

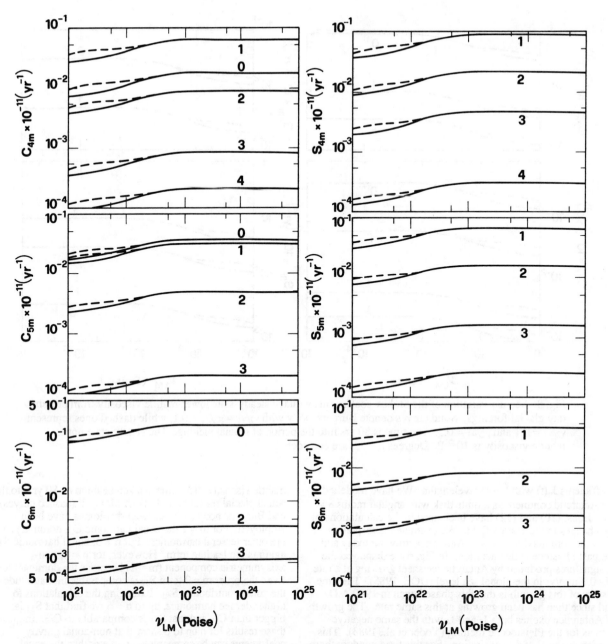

Fig. 9. Same as for Fig. 8. Degrees n = 4, 5, 6 are used.

geoid anomalies. It is important to separate out the relative contributions to the temporal variations of the geopotential made by the present cryospheric forcings, and that due to the last ice age, particularly the potential forcing from the Barents Sea (Peltier, 1988b). By means of transient viscoelastic modelling we have shown that the long-wavelength components of the geopotential are sensitive to current glacial discharges and also to the growth of the Antarctic ice sheet occurring today. Our results for the higher zonal harmonics reveal that Antarctica's mass disequilibrium may conceivably play an important role. Continuous monitoring of satellite orbits will be very useful for observing changes in the geopotential. Orbital inclinations different from LAGEOS-I will also shed light on the azimuthal dependence of the perturbed gravity field and may eventually be employed to put bounds on the amount of lateral variations in the viscosity. Progress in these areas depends crucially on the availability of future satellite missions, such as the forthcoming LAGEOS-II mission, devoted entirely to geodynamics, as LAGEOS-I has been doing during the past decade.

Acknowledgements. This research has been supported by CNR/PSN 86/060, NASA grant NAG 5-770 and NSF grant EAR-8511200. We thank Kari L. Rabie, Jill Borofka and Anne Boyd for preparing this paper.

References

Anderson, D.L. and J.B. Minster, The frequency dependence of Q in the earth and implications for mantle rheology and Chandler wobble, Geophys. J.R. astr. Soc., 58, 431-440, 1979.

Cathles, L.M., III, The Viscosity of the Earth's Mantle, Princeton University Press, Princeton, N.J., 1975.

Chao, B.F., O'Connor, W.P., Chang, A.T.C., Hall, D.K. and J.L. Foster, Snow load effect on the Earth's rotation and gravitational field, 1979-1985, J. Geophys. Res, 92, 9415-9422, 1987.

Dickman, S.R., Secular trend on the Earth's rotation pole: consideration of motion of the latitude observations, Geophys. J. Roy. Astron. Soc., 51, 229-244, 1977.

D.O.E. Report, Glaciers, ice sheets and sea level: effect of CO_2 induced climatic change, publ. by U.S. Dept. of Energy, Office of Energy Research, Wahington, D.C., DOE/ER/60 235-1, Sept., 1985.

Farrell, W.E., Deformation of the earth by surface loads, Rev. Geophys., 10, 761-797, 1972.

Gasperini, P., R. Sabadini and D.A. Yuen, Excitation of the Earth's rotational axis by recent glacial discharges, Geophys. Res. Lett., 13, 533-536, 1986.

Hager, B.H., R.W. Clayton, M.A. Richards, R.P. Comer and A.M. Dziewonski, Lower mantle heterogeneity, dynamic topography and the geoid, Nature, 313, 541-545, 1985.

Lambeck, K., The Earth's Variable Rotation: Geophysical Causes and Consequences, Cambridge Univ. Press, New York, 1980.

Lambeck, K. and S.M. Nakiboglu, Long-period Love numbers and their frequency dependence due to dispersion effects, Geophys. Res. Lett., 10, 857-860, 1983.

Meier, M.F., Contribution of small glaciers to global sea level, Science, 226, 1418-1421, 1984.

Merriam, J.B., LAGEOS and UT measurements of long-period earth tides and mantle Q, J. Geophys. Res., 90, 9423-9430, 1985.

Nakada, M. and K. Lambeck, Non-uniqueness of lithospheric thickness estimates based on glacial rebound data along the east coast of North America, in Mathematical Geophysics, ed. by N. Vlaar et al., pp 347-362, D. Reidel Publish. Co., Dordrecht, Netherlands, 1988.

O'Connell, R.J., Pleistocene glaciation and the viscosity of the lower mantle, Geophys. J.R. astr. Soc., 23, 299-327, 1971.

Peltier, W.R. and J.T. Andrews, Glacial-isostatic adjustment-I. The forward problem, Geophys. J.R. astr. Soc., 46, 605-646, 1976.

Peltier, W.R., Constraint on deep mantle viscosity from LAGEOS/acceleration data, Nature, 304, 434-436, 1983.

Peltier, W.R., New constraints on transient lower mantle rheology and internal mantle buoyancy from glacial rebound data, Nature, 318, 614-617, 1985.

Peltier, W.R., Lithospheric thickness, Antarctic deglaciation history, and ocean basin discretization effects in a global model of postglacial sea level change, in Mathematical Geophysics, ed. by N. Vlaar et al., pp 325-346, D. Reidel Publish. Co., Dordrecht, Netherlands, 1988a.

Peltier, W.R., Global sea level and earth rotation, Science, 240, 895-900, 1988b.

Rubincam, D.P., Postglacial rebound observed by LAGEOS and the effective viscosity of the lower mantle, J. Geophys. Res., 89, 1077-1088, 1984.

Sabadini, R., D.A. Yuen and E. Boschi, Polar wandering and the forced responses of a rotating, multi-layered viscoelastic planet, J. Geophys. Res., 87, 2885-2903, 1982.

Sabadini, R., D.A. Yuen and P. Gasperini, The effects of transient rheology on the interpretation of lower mantle viscosity, Geophys. Res. Lett., 12, 361-364, 1985a.

Sabadini, R., D.A. Yuen and R. Widmer, Constraints on short-term mantle rheology from the J_2 observation and the dispersion of the 18.6 yr. Love number, Phys. Earth Planet. Int., 38, 235-249, 1985b.

Sabadini, R., Yuen, D.A. and P. Gasperini, Mantle rheology and satellite signatures from present-day glacial forcings, J. Geophys. Res, 93, 437-447, 1988.

Smith, B.K. and F.O. Carpenter, Transient creep in orthosilicates, Phys. Earth Planet. Int., 49, 314-324, 1987.

Yoder, C.F., J.G. Williams, J.O. Dickey, B.E. Schutz, R.J Eanes and B.D. Tapley, Secular variation of the Earth's gravitational harmonic J_2 coefficient from LAGEOS and nontidal acceleration of Earth rotation, Nature, 303, 747-762, 1983.

Yoder, C.F. and E.R. Ivins, Changes in Earth's gravity field from Pleistocene deglaciation and present-day glacial melting, E.O.S., Vol. 66, No. 18, 245, 1985.

Yuen, D.A. and W.R. Peltier, Normal modes of the viscoelastic earth, Geophys. J.R. astr. Soc., 69, 495-526, 1982.

Yuen, D.A., R. Sabadini and E.V. Boschi, Viscosity of the lower mantle as inferred from rotational data, J. Geophys. Res., 87, 10 745-10 762, 1982.

Yuen, D.A. and R. Sabadini, Viscosity stratification of the lower mantle as infered by the J_2 observation, Ann. Geophys., 3, 647-654, 1985.

Yuen, D.A., R. Sabadini, P. Gasperini and E.V. Boschi, On transient rheology and glacial isostasy, J. Geophys. Res., 91, 11 420-11 438, 1986.

Yuen, D.A., Gasperini, P., Sabadini, R. and E. Boschi, Azimuthal dependence in the gravity field induced by recent and past cryospheric forcings, Geophys. Res. Lett., 14, 812-815, 1987.

MIGRATION OF VERTICAL DEFORMATIONS AND COUPLING OF ISLAND ARC PLATE AND SUBDUCTING PLATE

Satoshi Miura[1], Hiroshi Ishii[2] and Akio Takagi[1]

Abstract. Methods have been devised for analyzing vertical land deformation utilizing data of leveling survey. Two dimensional Chebychev functions for space domain and Akima's functions for time domain are employed for the analyses. The methods make possible to investigate continuous movements in time and filtered movements in space. By applying the methods, vertical deformations are investigated for the two areas, the Izu Peninsula and the northeastern Japan. Characteristics of uplift occurring in the northeastern area of the Izu Peninsula become clear and slow migration of an uplift peak with a velocity of about 10km/year is found.

In the case of the northeastern Japan numerical experiments are also performed by the use of the finite element method. In the computation the upper crust is assumed to be elastic, and the lower crust and the upper mantle, viscoelastic, taking account of the fact that most of the inland microearthquakes occur only in the upper crust. Deformations and stress accumulation caused by a sinking Pacific plate are computed. Computed deformations are compared with vertical deformations analyzed by the above methods and coupling of the island arc plate and subducting plate are investigated. It is revealed that in the northeastern Japan the Pacific ocean side is subsiding and the Japan sea side is uplifting.

Precursory uplift preceding the 1983 Japan sea earthquake (M=7.7) is found. It is also found that the Aseismic Front is interpreted as an end point of coupling area between the subducting plate and the continental plate overlain by the island arc starting from the trench.

Introduction

Data obtained from leveling surveys are very useful for investigating vertical deformations.

---

[1]Faculty of Science, Tohoku University, Sendai, Japan
[2]Earthquake Research Institute, University of Tokyo, Tokyo, Japan

Copyright 1989 by
International Union of Geodesy and Geophysics and American Geophysical Union.

Repeated surveys will provide knowledge of temporal and spatial variations in the deformations.

Kato and Kasahara (1977) proposed a method to represent space-time variations of vertical crustal movement along one leveling route by utilizing Spline function interpolation to both space and time domain. They successfully applied the method to a leveling route near the source area of the Niigata earthquake (1964, M=7.5). and clarified a precursory vertical crustal movement found by Tsubokawa et al. (1966) and Dambara (1973).

Vanicek et al. (1979) employed an algebraic polynomial to represent vertical movement as a function of position and time and revealed the characteristics of vertical crustal deformations in southern California. But the results were rather unstable because of the use of polynomial interpolation. Ishii et al. (1981) employed two dimensional Chebychev functions to express three dimensional vertical movements, and made possible a filtering operation in the space domain by selecting adequate order of the functions.

We devised a new method to obtain temporal and spatial variations of vertical movements from given leveling data. Applying the method to two areas of geophysical interest, vertical movements of the Izu Peninsula (Fig.2) and the northeastern Japan (Fig.7) are clarified.

Earthquake swarms sometimes occur off the Izu Peninsula whose northeast part is continuously uplifting with the rate of about 1.6 cm/year. In northeastern Japan, the Pacific Plate is sinking under the Eurasian Plate. A double plane seismic structure of microearthquakes was found and the upper plane coincides with the upper boundary of the sinking Pacific Plate (Takagi et al., 1977). Sato (1980) studied seismotectonics in the northeastern Japan Arc by using the Finite Element Method (FEM) applied to a viscoelastic medium. We have performed numerical experiments applying the program developed by him and investigated a relationship between vertical movements in the northeastern Japan and plate interaction.

Leveling Analysis Method

In this study, we devised a new method to obtain spatial and temporal variations of vertical

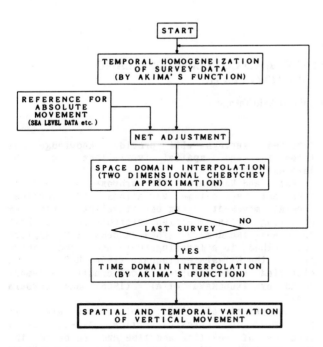

Fig. 1. Flow chart showing the process of leveling data analysis.

crustal movements from leveling data. Fig.1 is a flow chart showing the analysis procedure. Leveling surveys in Japan were started by the Military Land Survey at the end of the last century and continued by the Geographical Survey Institute (G.S.I.) after World War II. We use leveling data published by the G.S.I. in this paper. Usually, raw leveling survey data need preprocessing to correct for noise (e.g. land movements due to pumping of ground water), lost bench marks and so on. Large local elevation changes must be rejected because they do not conform with tectonic deformation.

In general, a large leveling survey (say, several hundred kilometers) is completed after two or three years. Time adjustment of the leveling data at each bench mark is necessary to relate the elevation changes at a common epoch. To do this adjustment, time series of elevation data at each bench mark are interpolated with straight lines or polynomial functions and the elevation changes at any time can be easily computed. Spline functions are often used for this kind of work, but spline function interpolation sometimes shows distorted curves different from the original data. Akima's function seems to be more stable for many cases (Akima, 1970). The Akima's function is often employed for automatic computer contouring of map and can draw natural and smooth curves.

Next, we proceed to the net adjustment. Since the raw survey data include closure errors, they must be corrected. A commonly used technique is to distribute the misclosure along a closed loop in proportion to the distance from the starting point. Informations about absolute vertical land movements at one or more reference points are needed from another data source (e.g. tide gauge data).

The next step is to calculate the elevation change at each grid point for each epoch from the adjusted leveling data and this procedure is repeated for every releveling.

For the case of the northeastern Japan, we adopted the method proposed by Briggs (1974). The method is often used for automatic machine contouring of data sampled at arbitrary intervals such as gravity surveys, geomagnetic surveys and so on.

We calculated Chebychev approximation function by utilizing Chebychev samplings obtained from interpolation of the values at the grid points. Therefore, orthogonality is valid and the Chebychev polynomial coefficients computed from the data are independent on the degree employed for the approximation. Ishii et al. (1978) applied Chebychev functions for analyses of crustal movements data and showed the effectiveness of their filtering operation.

For the case of the Izu Peninsula, we have adopted the Chebychev function for determining values at interpolation points on regular grids from observed data at irregularly placed bench marks. Chebychev polynomial converges faster than the other algebraic polynomial functions and shows better approximation with less order of polynomials. Let z_i be the elevation change at the i-th point (x_i, y_i), where $i=1,2,\ldots,N$ and N is the total number of points in the network. By using the Chebychev functions

$$T_n(x) = \cos\{n \cdot \arccos(x)\}, \quad n=1,2,\ldots, \quad (1)$$

a two dimensional approximation function $z(x,y)$ can be written as

$$z(x,y) = \sum_{k=0}^{m_x} \sum_{l=0}^{m_y} a_{kl} T_k(x) T_l(y), \quad (2)$$

where $m_x \cdot m_y < N$. We can determine the coefficients a_{kl} so as to minimize the summation of squares of deviations,

$$e_i = z_i - z(x_i, y_i). \quad (3)$$

Once the coefficients are determined, we can obtain values of z at arbitrary points (x,y) by computing the values of the approximation function $z(x,y)$. The coefficients, a_{kl} are calculated from the following simultaneous equations,

$$\sum_{k=0}^{m_x}\sum_{l=0}^{m_y} a_{kl}S_{klmn}=T_{mn}, \quad (4)$$

where $m=0,1,\ldots,m_x$, $n=0,1,\ldots,m_y$,

$$T_{mn}=\sum_{i=1}^{N} z_i T_m(x_i)T_n(y_i) \quad (5)$$

and

$$S_{klmn}=\sum_{i=1}^{N} T_k(x_i)T_l(y_i)T_m(x_i)T_n(y_i). \quad (6)$$

We employ this new method for the case of the Izu Peninsula.

Ishii et al. (1981) successfully used two dimensional Chebychev approximation functions for the analysis of vertical crustal movements in the Tohoku district. We can extract vertical movements of arbitrary spatial wave length easily by selecting adequate orders of coefficients of the Chebychev polynomial function. The process described above must be repeated for the number of surveys.

Finally, the elevation changes obtained at every grid point are summed up and then interpolated in the time domain throughout the survey epoch. We again use the Akima's function for this time domain interpolation. The procedure described above is applicable only for completely releveled networks. After all these procedures, we obtain the spatial and temporal variation of vertical crustal movement.

Vertical Movement of Izu Peninsula

Leveling routes of the Izu Peninsula located in the south Kanto district are drawn in Fig.2 (solid triangles show tidal stations). The lines are composed of both the first and the second order levelings. The analyzed period is from 1980 to 1986 and the survey was repeated every year. The vertical movements obtained by applying the method explained in the previous section, are presented in Fig.3. The UCHIURA tidal station is employed as a fixed point. Accumulated vertical displacements from August of 1980 are shown by contours for every year. The uplifting located at the northeast of the peninsula is easily recognized. The uplifting range becomes clear with the lapse of time; the center of the uplifting area with a radius of about 14 km is situated at the northeast of the peninsula. Though the uplifting area extends to the sea, the uplift off shore is not known for lack of observations. The maximum rate of uplift shown in Fig.3 is about 1.6 cm/year.

Next we will investigate the time variation of vertical deformations along the east coast

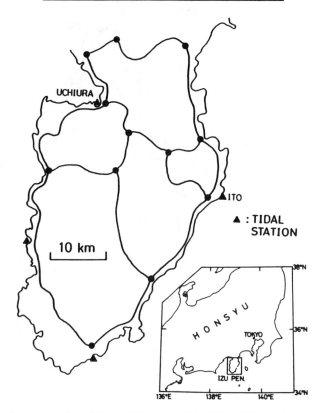

Fig. 2. Leveling routes in the Izu Peninsula.

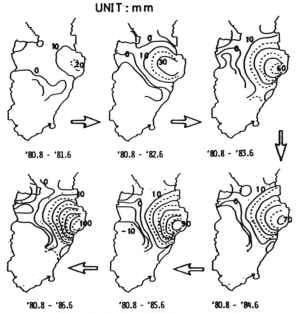

Fig. 3. Accumulated vertical displacements in the Izu Peninsula from August, 1980 to June, 1986.

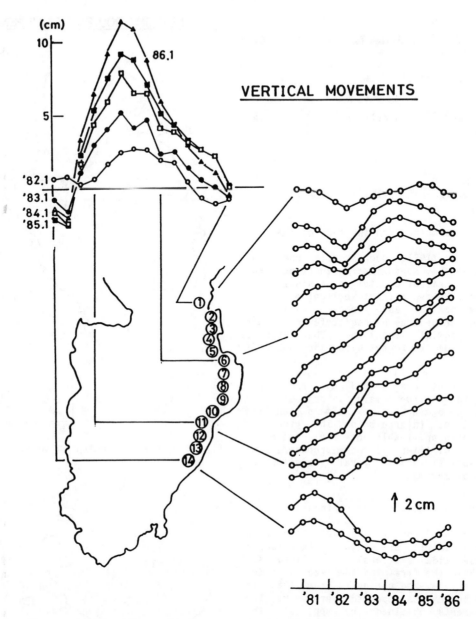

Fig. 4. Time variation of vertical movements on the route through uplift area along east coast in the Izu Peninsula.

illustrated in detail on Fig.4. Both temporal and spatial variations along the coast are evident. Most of the coast is being continuously uplifted and bench marks (B.M.) numbers 7 to 10 show especially large movements. The maximum uplift reached about 12 cm at B.M. 9 during the six years period. The temporal variation of B.M. 9 indicates very well coincidence with the temporal variation of the vertical movement obtained from the sea level data at Ito. The details will be reported elsewhere.

Slow Migration of Vertical Movements

In this section we will discuss the slow migration of the uplift peak in the Izu Peninsula. Fig.5 presents the time history of yearly vertical movements obtained by shifting the window by half

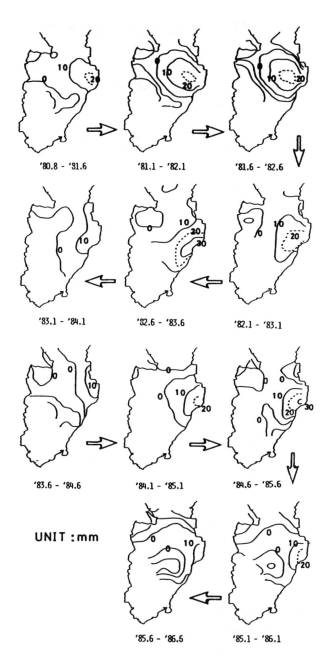

Fig. 5. Time variation of one year vertical movements obtained by shifting half a year from 1980.

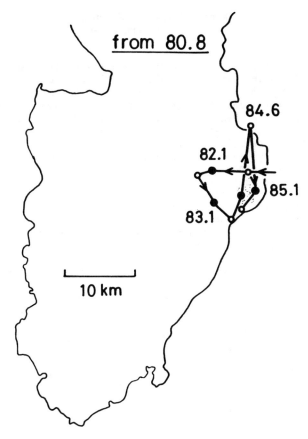

Fig. 6. Migration of uplift peak.

a year from 1980; the characteristics of uplift time variation are revealed. It is found that a peak of the uplift pattern varies with the lapse of time. The uplift peak sometimes moves from the land to the sea. In the case of the periods '83.1 - '84.1, '83.6 - '84.6 and '85.6 - '86.6 the peak is located in the sea area. From Fig.5 we can draw a vector diagram of the migration of the uplift peak as shown in Fig.6. The raw survey data along the leveling routes published by G.S.I. (1982, 1983, 1984, 1985, 1986, 1987) indicate the variation of the uplift peak location in time by itself. Therefore, this phenomenon is not an artifact of the data processing. The Chebychev approximation method is effective for objectively drawing the contours of vertical movements from data observed along the routes. On the average a speed of the migration is about 10 km/year.

Ishii et al. (1978, 1980) found a strain migrating in the northwest direction with a velocity of about 40 km/year in the northeastern Japan arc. Kasahara (1979) investigated migration of crustal deformation and pointed out some movements with velocity of several tens km/year. The mechanism of the migration of uplift peak is left to be elucidated. Sometimes earthquake swarms occurred in and off the Izu Peninsula (Ishida, 1987). However, the relationship between the uplift and occurrence of the swarms is not yet clear.

Vertical Crustal Movement of Northeastern Japan

Dambara (1971) compiled precise leveling data during the period from 1895 to 1965 and obtained a contour map of vertical movement for all of Japan. He concluded that the observed large subsidence near the Pacific Ocean coast of Hokkaido and northeastern Japan can be explained by a dragging force caused by the sinking plate. Kato (1979) investigated the vertical crustal movements between 1900 and 1975 in northeastern Japan (Tohoku district). He compared observed vertical movements with the results of the finite element calculation and concluded that a optimum model including tangential displacements at the plate boundary and downward ones around the boundary beneath the island part can explain the observed vertical movements.

Precise leveling surveys have been conducted by the Geographical Survey Institute at about ten year intervals for the last thirty years and at about five year intervals since the 1970's based on the national project of earthquake prediction. In this study we have adopted leveling data during the period 1956 to 1980 as data set homogeneous in time.

First order bench marks are set at spacing of about two kilometers along major national roads; however in this study we choose one bench mark out of every five, so the spacing of the data is about ten kilometers. Fig.7 shows the first order leveling routes in the Tohoku district and the locations of major tide gauge stations operated by the Geographical Survey Institute, the Japan Department of Japan. The surveys of the Tohoku district were adjusted to four epochs, '56, '66, '74 and '80 as explained before. We assumed the Hachinohe tide gauge station to be a fixed point. According to Kato and Tsumura (1979), the secular variation of the mean sea level at Hachinohe is the smallest among all shown stations.

Oblique projections and contour maps are computed and shown in Figs. 8 and 9 by using the results of spatial and temporal variations of the vertical crustal movement derived using the previously discussed procedures. Land movements of long wave length are demonstrated in Fig.8(b) by employing only low order coefficients (0-7 th order) out of 29 coefficients. The general features are a subsidence of the Pacific coast and an uplift of the north Japan Sea coast and the southern part of inland. The character of these features does not change with time. The last feature is regarded as a portion of the uplift zone found by Dambara (1971), which extends to the middle part of Honsyu, the main island of Japan. In Fig.9, the region where vertical movements are smaller than 2 cm is indicated by hatching, seems to shrink with time. The subsidence of the Pacific coast can be understood as a deformation of the island arc plate due to the dragging force of the

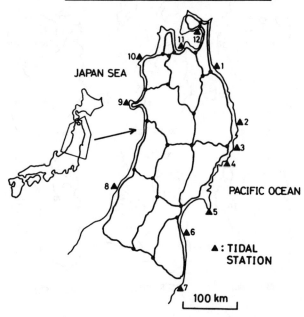

Fig. 7. Map showing first order leveling routes in the Tohoku district. Solid triangles with numerals indicate the location of major tide gauge stations operated by the Geographical Survey Institute, the Japan Meteorological Agency and the Hydrographical Department of Japan. 1:Hachinohe, 2:Miyako, 3:Kamaishi, 4:Ofunato, 5:Ayukawa, 6:Soma, 7:Onahama, 8:Nezugaseki, 9:Oga, 10:Fukaura, 11:Asamushi, 12:Ominato.

Pacific plate. We will discuss these results in the next section using numerical experiments.

As to the short wave length phenomena, there is a local subsidence around Aomori city located in the middle of the northern end of the analyzed area. This is regarded as an artificial deformation caused by pumping of ground water. There is another local subsidence area along the southern Japan Sea coast. This subsidence is coseismic one due to the Niigata earthquake (1964, M=7.5) which occurred very close to the coast. Though the phenomenon seems to accumulate gradually with time, it is an apparent variation because of the time domain interpolation. We should refer to only the amount of the elevation change between the two survey epochs, '56 and '66. Abe (1974) proposed a source model for this earthquake based on the analysis of body wave, surface wave and geodetic data. According to the model, the maximum coseismic subsidence along the nearest route reaches twenty centimeters. The agreement between the observed vertical movements and those calculated from the model is quite good.

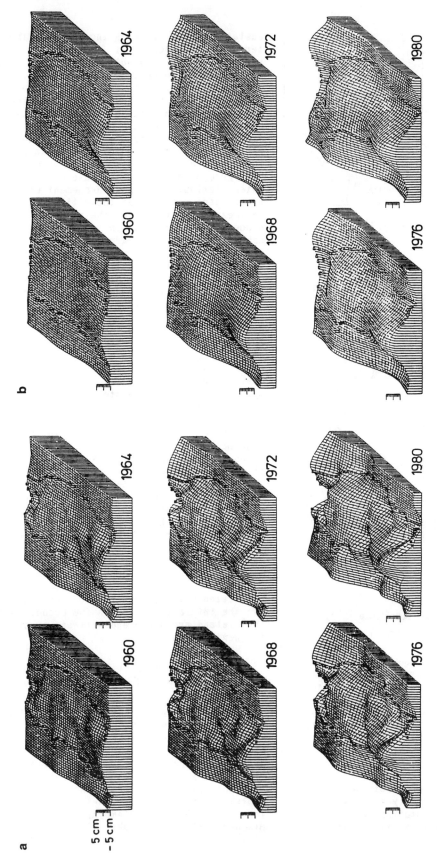

Fig. 8. Oblique projection maps showing the spatial and temporal variation of vertical land movement. Accumulated land movement from 1956 are plotted. Small squares represent coast lines. (a) 0-28 orders of the Chebychev approximation functions are used. (b) Only 0-7 orders are selected for plots. Long wave length movements are successfully extracted.

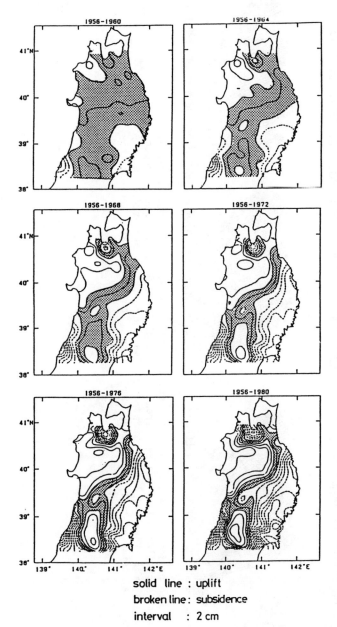

Fig. 9. Contour maps of accumulated vertical movement with time interval of four years, contour interval being 2 cm, where 0-28 orders of the Chebychev polynomials are used. Solid lines indicate uplift and broken lines, subsidence. Hatched areas indicate the region where vertical movements are smaller than 2 cm in absolute value.

In Fig.10, vertical movements obtained from the secular variation of the yearly mean sea level observed at some tide gauge stations are compared with the results of this study. Though the eustatic water rise is included in the secular term of the sea level change, the amount is relatively small, about 1 mm/year, in comparison with another terms such as coseismic, tectonic, and local ones. Here, we take no account of the eustatic water rise and compare only the general features of both data sets. At Nezugaseki, a large steplike subsidence due to the coseismic displacement associated with the Niigata earthquake can be seen in the tide data. However, the vertical movements derived from our method show gradual subsidence because of the time domain interpolation. The good agreement of these two independent data set at these tide stations except for Nezugaseki indicates the propriety of the data processing proposed in this study and the assumption of the fixed point.

Fig. 11(b) shows vertical crustal movements along cross sections indicated in Fig. 11(a) taken so as to be parallel to the direction of the subduction of the oceanic plate. Note that the large subsidence along the southern Japan Sea coast observed in cross sections from H-H' through K-K' is the effect caused by the Niigata earthquake. Accumulated vertical movements starting from 1956 are plotted for every two years. Some characteristics mentioned above are obvious again, including the acceleration of uplift around the western end in the cross sections from B-B' to E-E'. This region is close to the source area of the 1983 Japan Sea earthquake (M=7.7) as shown in Fig.12. The abnormal land uplifts are considered to be a precursor of the earthquake (see also Ishii et al., 1986, Fig.15).

It must be also noted that all cross sections have one or two nodal points, at which vertical movement is almost zero and these do not move with time. It can be clearly seen in Fig. 9 as mentioned before. This means that the analyzed region is distinctly divided into uplifting and subsiding areas and the mode of long term vertical movement has not changed throughout the period from 1956 to 1980. This observed fact may be associated with the crustal deformation generated by the interaction between the oceanic plate and the island arc plate and will be investigated in the next section.

Coupling of Island Arc Plate
and Subducting Plate

We next interpret observations as described in the preceding section from a plate tectonic viewpoint using numerical simulations. The finite element method (FEM) is useful computations of the deformation and the stress in the case of a medium with complex shape, inhomogeneities, inelasticity, and thermal stress because these cases cannot be analytically solved. It has been recently used for solid earth problems by many authors (e.g. Bischke, 1974; Shimazaki, 1974; Kato, 1979; Sato

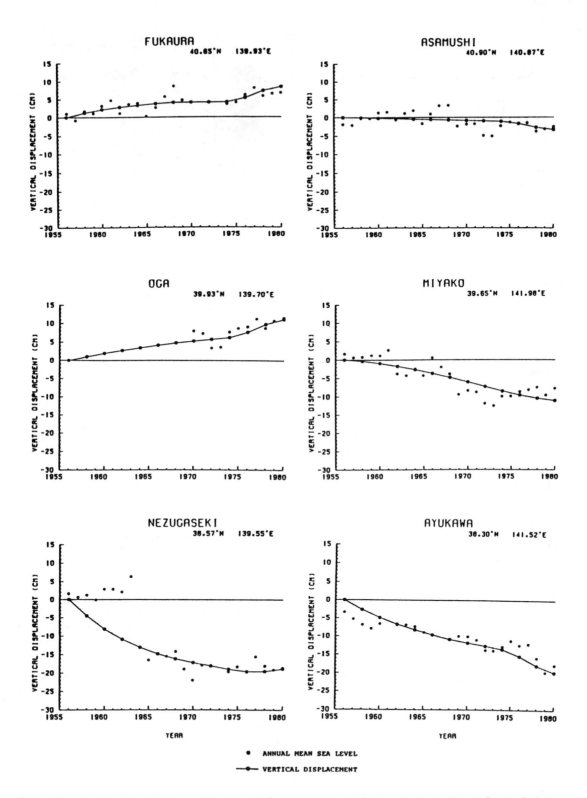

Fig. 10. Comparison between the vertical movements at tide gauge stations derived from the secular variation of the yearly mean sea level (solid circles) and that calculated by using the method proposed in this study (open circles with lines).

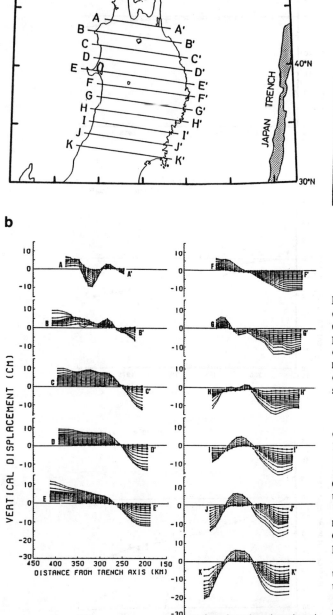

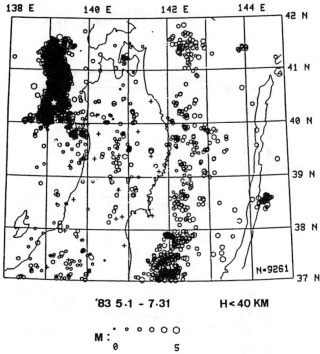

Fig. 11. (a) Cross sections along which vertical land movements are plotted. The direction of each cross section is taken so as to be almost parallel to the direction of subduction of the oceanic plate. (b) Accumulated vertical movements for every two years starting from 1956 along cross sections indicated in (a).

Fig. 12. Map showing the distribution of epicenters of microearthquakes obtained from the observation network of Tohoku University for the period from May 1 to July 31, 1983. Epicenters concentrated in the northwestern corner of the map are the aftershocks of the 1983 Japan Sea earthquake (M=7.7). The epicenter of the main shock is indicated by an open star.

et al., 1981; Hashimoto, 1982, 1984; Goto et al., 1985 etc.). Sato et al. (1981) used the method to solve two dimensional viscoelastic problems, investigated the time dependent stress distribution and deformation in the crust and upper mantle, and discussed the seismotectonics in the northeastern Japan arc. We apply the same method to compute the vertical crustal movement due to the interaction between the island arc plate and the subducting plate.

Fig.13 shows the model of the crust and the upper mantle beneath the northeastern Japan arc used in our simulation. A curved boundary along the right end of the model is presumed to be the plate boundary. The crust is divided into three sections, U.C. (upper crust), L.C.1, and L.C.2 (lower crust) and the upper mantle, into three sections, U.M.1, U.M.2 and U.M.3. Physical constants of each section are determined by using the results from many studies and listed in Table 1. The upper crust is assumed to be elastic, and the lower crust and the upper mantle, to be viscoelastic, taking account that most of the inland microearthquakes occur only in the upper

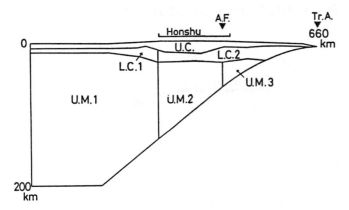

Fig. 13. Two dimensional structure model of the island arc plate employed for the FEM simulation. U.C. represents the upper crust, L.C.1 and L.C.2, the lower crust, and U.M.1, U.M.2 and U.M.3, the upper mantle, respectively. The island arc plate is in contact with the oceanic plate descending from the trench axis, denoted as Tr.A, along the curved border. Horizontal distance and depth are measured from the upper left corner.

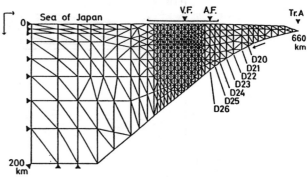

Fig. 14. Two dimensional finite element mesh used in the computation. Solid triangles indicate fixed nodes for boundary conditions. Tangential downward displacement were given along the plate boundary from the tip of the wedge (trench axis, Tr.A) down to the node indicated as D20,...,D25, and D26 taking account of dragging force generated by the coupling between the two plates. Other boundary nodes are under stress free conditions. Horizontal distance and depth are measured from the upper left corner.

crust (Takagi et al., 1977). This phenomenon was supported by the study on the stress distribution in the crust and upper mantle (Sato, 1980). The sectioning of the model and the values of the constants are adopted from work dealing with the velocity and Q structures of seismic waves (e.g. Yoshii and Asano, 1972), and postseismic rebound (Thatcher and Rundle, 1979).

The mesh for the FEM is shown in Fig.14. The mesh division is finer for the inland portion than for the surrounding portions. Fig.14 also shows the assumed boundary conditions. The left and the bottom edges are fixed area though the inland arc plate and the mantle extend beyond this boundary. The surface and the deeper part of the plate boundary are assumed to be stress free. The latter assumption is based on the speculation that at the deeper part of the plate boundary, the oceanic plate is dragging down without any interaction between two plates and they might be deformed ductilely. The stress free condition dose not stand for the exact state but approximated one.

TABLE 1. Physical Constants of Medium for the FEM Simulation.

Medium	Young's Modulus (10^{11}Pa)	Poisson's Ratio	Viscosity (10^{19}Pa·s)
U.C.	0.808	0.226	∞
L.C.1	1.21	0.258	4.0
L.C.2	1.77	0.258	4.0
U.M.1	1.77	0.273	4.0
U.M.2	1.50	0.273	4.0
U.M.3	1.83	0.245	4.0

The interaction between the two plates is represented by the tangential downward displacement with a constant velocity, 3 cm/year, for the nodes along the plate boundary from the tip of the wedge, the trench axis. We have considered cases with different boundary conditions. Symbols D20,..., D25 and D26 in Fig.14 indicate the nodes of the deepest end to which the tangential displacements are given; the result for each boundary condition is shown in Fig.15. The horizontal axis represents the distance from the trench axis and vertical displacements are plotted for every four years. It is clearly seen that the subsiding area extends to the inland region as the coupling region becomes deeper and that uplifting and subsiding region are divided by a nodal point.

The position of the nodal point seems to reflect the position of the end point of the coupling area very sensitively. We now compare the computed positions with the analyzed ones in Fig.11 and find similarities regarding the nodal point and the pattern of displacement.

Comparisons between the computed and observed results are presented in Fig.16(a) indicating the length of surface projection of the coupling area from the trench axis versus the nodal point distance from the trench axis. By reading values of the ordinate of the plots from A to K, the projected lengths of the coupling area are

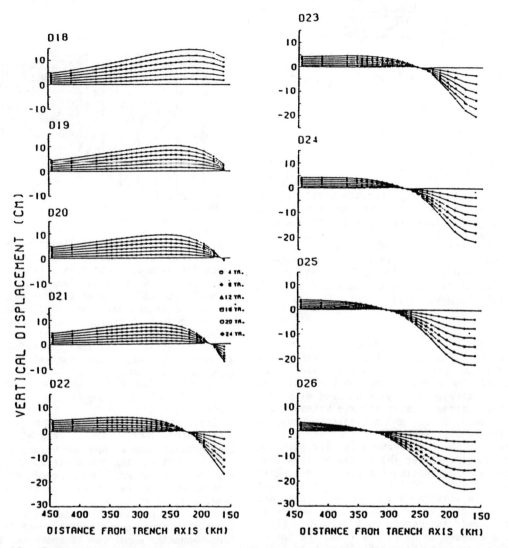

Fig. 15. Computed vertical displacement at the surface of the model plotted for every four years for various boundary conditions.

obtained and plotted as hatched bars in Fig.16(b), where the Aseismic Front is indicated as A.F. It is noted that the Aseismic Front almost corresponds with the location of the projected western tip of the coupling area. The Aseismic Front was originally proposed by Yoshii (1975) for the western boundary of a seismic belt, where many earthquakes were occurring at the depth from 40 to 60km, extending to the trench axis. The Front is also a western border of source areas of big inter-plate earthquakes. Therefore, the Front can be interpreted as the boundary of coupling area between the two plates using not only seismological but also geodetic data.

Though the numerical simulation employed in this study is two dimensional, our model represents actual crustal movement of this region with sufficient accuracy because the oceanic plate beneath northeastern Japan is approximately considered to be subducting in the same manner everywhere. Since we investigated by using two dimensional model along the direction of the plate subduction, the results are expected not to be very different from a three dimensional computation. Considering the rapid progress of capacity of computers, executions of huge programs will be getting faster and cheaper. Three dimensional simulation with boundary conditions

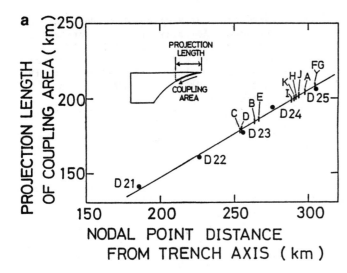

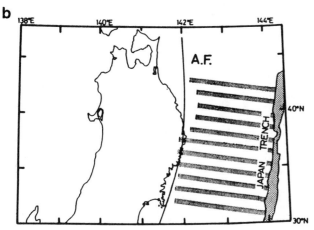

Fig. 16. (a) Length of the surface projection of the coupling area measured from trench axis versus the nodal point distance from trench axis. Solid circles are plotted for the result computed from different boundary conditions. Short bars with capitals indicate the location of the nodal point read from Fig.11 for each cross section. (b) Surface projection of coupling area estimated from this study (hatched region). The curve denoted with A.F. is the Aseismic Front determined from seismological data.

representing more realistic geophysical situation will be presented in the near future.

Conclusions

We have devised a new method to reconstruct vertical crustal movements by the use of leveling data. Interpolations was carried out in the space and time domains utilizing the Chebychev approximation functions and the Akima's function, respectively. Discrete data sets in space and time derived from leveling surveys are transformed into continuous data set in space-time domain, making convenient the investigation of characteristics of land movements in detail. The advantages of the adoption of the Chebychev functions are not only the efficient approximation in space domain but also the easy operation of filtering procedure in space domain.

We have applied the method for two geophysically interesting region in Japan. One is the Izu Peninsula located in the middle of Japan, where the Philippine Sea plate is subducting beneath the continental plate and colliding with the Pacific Ocean plate, and the seismic activity is very high. Slow migration of an uplifted peak with a velocity of about 10 km/year was found in this region. Though the mechanism of this phenomena and its geophysical meaning are left for a further study, it should have some relation to the inelastic behavior of the crust as the strain migration found in northeastern Japan (Ishii et al., 1978, 1980).

The other is the Tohoku district, northeastern Japan, where the Pacific Ocean plate is subducting under the continental plate and the double seismic plane can be seen very clearly (Umino and Hasegawa, 1975), creating the typical island arc-trench system. We have performed numerical experiments to simulate the vertical land movements. We have assumed that the upper crust is elastic and the lower crust and the upper mantle viscoelastic. The observed land movements are well modeled by the numerical experiments. The coupling area between the continental and the oceanic plate is ascertained by the analyses and the western margin of the coupling area corresponds to the Aseismic Front. The significance of the Front from the viewpoint of plate tectonics has been clarified.

Acknowledgments. The authors express their sincere thanks to Prof. T. Hirasawa for his instructive suggestions and discussions. Thanks permission to use his FEM program and useful comments. The authors greatly appreciate valuable discussions with the staff of the Observation Center for Prediction of Earthquakes and Volcanic Eruptions, the Akita Geophysical Observatory, the Honjo Seismological Observatory, the Kitakami Seismological Observatory and the Sanriku Geophysical Observatory. The authors are also grateful to the referees who gave valuable comments.

References

Abe, K., Re-examination of the Fault Model for the Niigata Earthquake 1964., J. Phys. Earth, 23, 349-366, 1975.

Akima, H., A New Method of Interpolation and Smooth Fitting Based on Local Procedures, J. Assoc. Computing Machinery, 17, 589-602, 1970.

Bischke, R., A Model of Convergent Plate Margins Based on the Recent Tectonics of Shikoku, Japan, J. Geophys. Res., 79, 4845-4857, 1974.

Briggs, I., Machine Contouring Using Minimum Curvature, Geophys., 39, 39-48, 1974.

Dambara, T., Synthetic Vertical Movements in Japan during the Recent 70 Years (in Japanese), J. Geod. Soc. Japan, 17, 100-108, 1971.

Dambara, T., Crustal Movements before and after the Niigata Earthquake (in Japanese), Rep. Coord. Comm. Earthq. Predict., 9, 93-96, 1973.

Geographical Survey Institute, Crustal Movement in the Izu Peninsula (in Japanese), Rep. Coord. Comm. Earthq. Predict., 27, 155-174, 1982.

Geographical Survey Institute, Crustal Movement in the Izu Peninsula (in Japanese), Rep. Coord. Comm. Earthq. Predict., 29, 147-167, 1983.

Geographical Survey Institute, Crustal Movement in the Izu Peninsula (in Japanese), Rep. Coord. Comm. Earthq. Predict., 31, 229-245, 1984.

Geographical Survey Institute, Crustal Movement in the Izu Peninsula (in Japanese), Rep. Coord. Comm. Earthq. Predict., 33, 236-257, 1985.

Geographical Survey Institute, Crustal Movement in the Izu Peninsula (in Japanese), Rep. Coord. Comm. Earthq. Predict., 35, 235-248, 1986.

Geographical Survey Institute, Crustal Movement in the Izu Region (in Japanese), Rep. Coord. Comm. Earthq. Predict., 37, 224-242, 1987.

Goto, K., H. Hamaguchi and Z. Suzuki, Earthquake Generating Stress in a Descending Slab, Tectonophys., 112, 111-128, 1985.

Hashimoto, M., Numerical Modeling of the Three-dimensional Stress Field in Southwestern Japan, Tectonophys., 84, 247-266, 1982.

Hashimoto, M., Finite Element Modeling of Deformations of the Lithosphere at an Arc-Arc Junction: the Hokkaido Corner, Japan, J. Phys. Earth, 32, 373-398, 1984.

Ishida, M., Recent Seismic Activity in and around the Izu Peninsula (in Japanese), Proceedings of Earthquake Prediction Research Symposium, 51-60, 1987.

Ishii, H., T. Sato and A. Takagi, Characteristics of Strain Migration in the Northeastern Japanese Arc (1) -- Propagation Characteristics --, Sci. Rep. Tohoku Univ., Fifth Ser., Geophys., 25, 83-90, 1978.

Ishii, H., T. Sato and K. Tachibana, Observation of Crustal Movements at the Akita Geophysical Observatory (3) -- Application of Chebychev Approximation Function for Data Observed by Extensometer and Tiltmeter --, J. Geod. Soc. Japan, 24, 122-131, 1978.

Ishii, H., T. Sato and A. Takagi, Characteristics of Strain Migration in the Northeastern Japanese Arc (2) -- Amplitude Characteristics --, J. Geod. Soc. Japan, 26, 17-25, 1980.

Ishii, H., Y. Komukai and A. Takagi, Characteristics of Vertical Land Movement and Microearthquake Activity in the Northeastern Japan Arc, Tectonophys., 77, 213-231, 1981.

Ishii, H., S. Miura and A. Takagi, Large-scale Crustal Movements before and after the 1983 Japan Sea Earthquake, J. Phys. Earth, 34, Suppl., S159-174, 1986.

Kasahara, K., Migration of Crustal Deformation, Tectonophys., 521, 329-341, 1979.

Kato, T., Crustal Movements in the Tohoku District, Japan, During the Period 1900-1975, and Their Tectonic Implications, Tectonophys., 60, 141-167, 1979.

Kato, T. and K. Kasahara, The Time-Space Domain Presentation of Leveling Data, J. Phys. Earth, 25, 303-320, 1977.

Kato, T. and K. Tsumura, Vertical Land Movement in Japan as Deduced from Tidal Record, (1951-1978), Bull. Earthq. Res. Inst., 54, 559-628, 1979.

Sato, K., A Study on the Seismotectonics in the Northeastern Japan Arc (in Japanese), Master Thesis, Tohoku University, pp.163, 1980.

Sato, K., H. Ishii and A. Takagi, Characteristics of Crustal Stress and Crustal Movements in the Northeastern Japan Arc I: Based on the Computation Considering the Crustal Structure (in Japanese), Zisin (J. Seismol. Soc. Japan), 34, 551-563, 1981.

Seno, T., K. Shimazaki, P. Somerville, K. Sudo and T. Eguchi, Rupture Process of the Miyagi-oki, Japan Earthquake of June 12, 1978, Phys. Earth Planet. Inter., 23, 39-61, 1980.

Shimazaki, K., Preseismic Crustal Deformation Caused by an Underthrusting Oceanic Plate, in Eastern Hokkaido, Japan, Phys. Earth Planet. Inter., 8, 148-157, 1974.

Takagi, A., A. Hasegawa and N. Umino, Seismic Activity in the Northeastern Japan Arc, J. Phys. Earth, 25, Suppl., S95-S104, 1977.

Thatcher, W. and J. B. Rundle, A Model for the Earthquake Cycle in Underthrust Zones, J. Phys. Res., 84, 5540-5556, 1979.

Tsubokawa, I., Y. Ogawa and T. Hayashi, Crustal Movements before and after Niigata Earthquake, J. Geod. Soc. Japan, 10, 165-171, 1966.

Umino, N. and A. Hasegawa, On the Two-Layered Structure of Deep Seismic Plane In Northeastern Japan Arc (in Japanese), Zisin (J. Seismol. Soc. Japan), 28, 125-139, 1975.

Vanicek, P., M. R. Elliott and R. O. Castle, Four-dimensional Modeling of Recent Vertical Movements in the Area of the Southern California Uplift, Tectonophys., 52, 287-300, 1979.

Yoshii, T., Proposal of the "Aseismic Front" (in Japanese), Zisin (J. Seismol. Soc. Japan), 28, 365-367, 1975.

Yoshii, T. and S. Asano, Time-term analysis of explosion seismic data, J. Phys. Earth, 20, 47-57, 1972.